Rodolfo Andreani Sobrinho

Indices for monitoring Culex quinquefasciatus in swimming pools

Rodolfo Andreani Sobrinho

Indices for monitoring Culex quinquefasciatus in swimming pools

Culicidae infestation control for managers

ScienciaScripts

Imprint
Any brand names and product names mentioned in this book are subject to trademark, brand or patent protection and are trademarks or registered trademarks of their respective holders. The use of brand names, product names, common names, trade names, product descriptions etc. even without a particular marking in this work is in no way to be construed to mean that such names may be regarded as unrestricted in respect of trademark and brand protection legislation and could thus be used by anyone.

Cover image: www.ingimage.com

This book is a translation from the original published under ISBN 978-3-330-76234-3.

Publisher:
Sciencia Scripts
is a trademark of
Dodo Books Indian Ocean Ltd. and OmniScriptum S.R.L publishing group

120 High Road, East Finchley, London, N2 9ED, United Kingdom
Str. Armeneasca 28/1, office 1, Chisinau MD-2012, Republic of Moldova, Europe
Managing Directors: Ieva Konstantinova, Victoria Ursu
info@omniscriptum.com

Printed at: see last page
ISBN: 978-620-8-38759-4

DEDICATORY

TO MY BELOVED WIFE

CYNTHIA SANTOS ANDREANI,

WITHOUT WHOM I WOULDN'T HAVE DONE A TENTH OF THIS WORK, I DEDICATE IT TO YOU. YOUR STRENGTH AND KINDNESS MAKE ME LIVE.

IS MY WATER.

TO MY CHILDREN DANTE, AUGUSTO AND MARIANA,

I OFFER IT WITH ALL MY LOVE.

ARE MY OXYGEN, MY SUN AND MY BLOOD.

ACKNOWLEDGEMENTS

- To Prof Dr Delsio Natal, who despite being a bastion of national entomology, always found time, encouragement and attention for me through his generous guidance.

- To Dr Paulo Roberto Urbinatti, who acted practically as a co-supervisor, but not formally, which makes his detachment even more valuable.

- To the biologist Cinthia Santos Andreani, who accompanied me on all the field visits for her essential operational support and for all the laboratory help in counting and identifying the species.

- To the biologists Viviane Galhato, Nayara Menegasso and Luzinete Cabral de Oliveira for accompanying us at the start of the field work, for their support in taking photographs in the pools, for their support in laboratory identification and also for the little help during the course of the research.

- To the Zoonosis Control Management, Health Surveillance Department, Municipal Health Department, Santo André City Hall for allowing me to study for a Professional Master's Degree.

- To Dr Simone Rizzotti and administrative manager Andréa Semensato and leader Emerson Carlos Gregório for providing data on pest control and complaints, and to coordinator Antonio Carlos Cavalcanti for his solicitude.

- To retired employee Francisco Machado (Tico) for his help in making the resting boxes.

- To the Health Warehouse, especially Andrei dos Santos Bastos for providing the Styrofoam boxes.

- To my father-in-law Ededeus Santos, for painting the resting boxes.

- To my nephew Wilian Santos Viana, for his spatial referencing work and computer support.

- To Prof Dr Aristides Fernandes, for having taught the course on culicidae identification to the biologists mentioned above.

- To Drainage Manager Engineer Angélica Ferrini for providing data on the cleaning and weeding of the Bom Pastor reservoir, as well as institutional support. Many thanks to Drainage Management employees Laura Sueli Vieira de Melo and Rute Yuri Kishimoto for providing data on drainage and weeding at the Bom Pastor Reservoir.

- To Mr Edson Celeni and Mr Paulo Gustavo Hereny Gonçalves de Oliveira, Manager and Head, respectively, of Network and Reservoir Operations, responsible for the climatological information and the Annexes to this dissertation.

- To Mr Alceu Segamarchi Júnior, Superintendent of the Department of Water and Electric Energy - DAEE, for allowing access and carrying out the work at the RM 9 Reservoir, Faculty of Medicine, Valparaíso.

- To Engineer Ubirajara Tadeu Ferreira, Technical Director of the DAEE for his monitoring, guidance and technical information on the RM 9 Reservoir, Faculty of Medicine, Valparaíso.

- Edilson Souza do Nascimento, Rodrigo dos Santos Vitório and Manoel, the watchmen at the RM 9 Medical School Reservoir, for their promptness and kindness.

- To the driver leader Mauro Aparecido Neves for his professionalism and to the drivers who accompanied me with their help and dedication: José Moreira dos Santos, Mauro da Silva (Parça), José Laurindo (Papito), Donizete Souza, Antonio Alonso (Bolinha), João Donizetti Branco, Wagner Barbosa (Abelha), Benedito (Benê).

- To agronomists Antonio Senise Júnior and Luis Rogério Gonçalves from Sol Comércio for their support for my previous project and for their technical information on insecticides.

- To my classmates from the Professional Master's programme, especially my friend Engineer Marcelo Loutfi, with whom we did at least four academic projects together. Thanks to Fernando Toffo for his good wishes. I would also like to thank Engineer Arisnandes Antonio da Silva for his help in preparing the Product (Appendix G) of this dissertation.

- To the professors at the School of Public Health, for making an extraordinary contribution to my professional training. I would like to thank Prof Dr Leandro Giatti for his good wishes for my recovery.

- To the University of São Paulo, for opening the Professional Master's programme and giving me this opportunity.

- To all the staff at the FSP, especially José, Vera Sampaio and the others in the Library for clearing up many doubts and to Pedro Daniel for clarifying the administrative procedures.

- To all my family and everyone who has contributed to my personal and professional development.

- To veterinary medicine and animals.

- To God, for being alive.

SUMMARY

ANDREANI SOBRINHO, R. **Application of** Culex quinquefasciatus **infestation indices and recording of nuisance complaints in areas related to swimming pools in Santo André, SP.** 2015. 141 f. Dissertation (Master's) - School of Public Health, University of São Paulo, São Paulo, 2015.

Introduction: Culicidae (insects of the Order Diptera, Family Culicidae), commonly known as mosquitoes, are vectors of various pathogens of human and animal diseases and a nuisance to the general population. Resources launched with the aim of controlling urban pests using chemical compounds must be used as sparingly as possible in order to avoid environmental contamination and hinder the emergence of resistant strains of vectors and urban pests to the biocides used. **Objectives: To** establish Culex quinquefasciatus infestation rates based on the count of adult insects and their immature forms collected in swimming pools and to propose a tool for estimating the discomfort caused by bites to the surrounding population. **Materials and Methods:** The period for collecting mosquito specimens and data was from August 2014 to May 2015. Two swimming pools in the municipality of Santo André, the Bom Pastor Neighbourhood Pool and the Medical School Pool (Valparaíso), were chosen as the study area. This research proposes the construction of some infestation indices for the species under study based on two aspects: 1) - A diagnosis was made by counting mosquitoes in the area of the swimming pools. The adult specimens of this mosquito were counted and identified by vacuuming the shelters and vacuuming the resting boxes that were installed in the pool areas, and the immature forms (larvae and pupae) collected from the aquatic breeding sites in the pools were counted. 2) - Complaints from residents were analysed via the City Council's Service Information System, within a radius corresponding to the area of influence of each reservoir. **Results:** A total of 1,896 adult specimens were collected when vacuuming the shelters, 621 immature specimens and 67 adult specimens vacuumed from the resting boxes. The sex of the adult specimens was identified and males predominated. Three infestation indices (Density of Immatures per Litre, Index of Mosquitoes Aspirated per Minute and Index of Mosquitoes Aspirated from Resting Boxes) and an index of complaints (Index of Corresponding Complaints) were created and applied based on the sampling carried out, observing natural and anthropogenic variables and analysing complaints. **Conclusions:** By analysing seasonal aspects, the indices prove to be important tools for pest control interventions. The resting box method proved to be less productive than the other indices.

Keywords: mosquitoes, Culicidae, swimming pools, infestation rates, complaint rates.

Summary

CHAPTER 1

INTRODUCTION

1.1 THE PROBLEM OF FLOODING IN THE METROPOLITAN REGION OF SÃO PAULO

The São Paulo Metropolitan Region (RMSP), with 39 municipalities and a population of over 19 million inhabitants, which includes the municipality of Santo André, is a complex and fragile ecosystem. It is located in the south-east of Brazil and is the third largest urban conglomerate in the world, covering an area of 7944 Km^2 (Jacobi, 2013).

The RMSP's growth rate over the last few decades has been around 0.96 % per year and despite the slowdown in growth in recent years there has been high demographic pressure to occupy areas known as floodplains and consequent soil sealing, especially on the fringes of the urban area (Jacobi, 2013). The removal of vegetation and soil sealing in cities drastically reduces the infiltration of rainwater into the soil (Silvério, 2008, Giroldo, 2003).

The RMSP is located in the Alto Tietê Hydrographic Basin. In this basin, in addition to the Tietê River itself, the main watercourses are its tributaries: the Pinheiros, Aricanduva, Cabuçu-de-cima, Cabuçu-de-baixo, Itaquera and Tiquatira rivers. In addition to the rivers, the basin is home to the Billings and Guarapiranga reservoirs. The basin covers an area of 5500 square metres in the metropolitan region and is bordered to the south and southeast by the Serra do Mar, Taxaquara, Itatuba and Taquaxiara mountains, to the north by the Cantareira, Bananal and Itaqui mountains and to the east by the Itapeti mountain range and, due to this mountainous boundary, its only outlet is the Tietê River. This location leads to low water flow velocities, and when there is heavy rainfall, the flow capacity of the river and its tributaries is exceeded, causing flooding (Giroldo, 2003).

Jacobi (2013) received oral information from technicians at the Institute for Technological Research (IPT) that more than 500 points of flooding were estimated in the RMSP, affecting the quality of urban life in 2012.

The municipality of Santo André was officially founded on 8 April 1553. With an area of 174.38 square kilometres, it is located in the Greater ABC region (Metropolitan Region of São Paulo), 18 kilometres from the capital. The city is strategic for the logistics sector, as it is located in the country's main economic centre, close to some of the main state and federal highways, which give access to the Port of Santos and the Cumbica and Congonhas airports (PMSA, 2013).

According to the latest census, published in 2011, Santo André has 678,486 inhabitants. In the same year, the GDP (Gross Domestic Product) was R$16.9 billion, the 29th highest in the country and the 10th highest among the cities in the state of São Paulo. The 2013 budget is R$2.4 billion (PMSA, 2013).

As a result of urban development since the 1950s, there has been a notable increase in the number of vehicles using fossil fuels, generating intense air pollution. The regional climate has suffered impacts that are currently manifested in the form of heat islands that can raise the ambient temperature by up to 10°C in certain parts of the urban environment. The geographical position of the metropolitan region is subject to the advance of breezes from the coast and their encounter with the heat islands formed over the large paved area, a phenomenon that causes storms with high rainfall and generates serious flooding in the urban environment (Giroldo, 2003; Silvério, 2008; Nobre,

2010).

The Meteorological Station of the Institute of Astronomy and Geophysics of the University of São Paulo (IAG-USP) has demonstrated through meteorological measurements in the Metropolitan Region of Greater São Paulo since 1933 that there has been an increase in the frequency and magnitude of rainfall over the last 80 years and is associated with air pollution and increased rainfall in the dry season, due to changes in the micro-physics of clouds (Dias et al. 2012).

In addition to these issues, the increase in the population living in unsuitable housing conditions, with limited solid waste collection, combined with deficiencies in environmental education, result in the disposal of this waste directly on the ground or in water bodies, causing drainage systems such as culverts and manholes to clog, further aggravating the flooding situation (Besen et al., 2010; Jacobi, 2013). Freitas et al. (2012) demonstrated in their study how social conditions and environmental education have a direct impact on socio-environmental vulnerability by analysing and comparing environmental tragedies in Brazil, Chile and Haiti, concluding that many lives and financial resources could be saved if there were greater investment in prevention and education, regardless of the magnitude of the incident.

Notwithstanding education, the population's own increase in consumption and the demographic explosion, there is a considerable increase in the production of solid waste combined with market rules such as the use of disposable packaging and programmed obsolescence, which further contribute to aggravating the situation of increasing waste and the consequent collapse of drainage systems (Besen et al. 2010).

Due to demographic pressure, the search for housing has a deleterious effect on water source regions, resulting in the loss of riparian forest, reduced vegetation cover, soil compaction, increased erosion rates, among other aggravating factors. As a result, there is an increase in the silting up of water reserves and an increased risk of landslides (Maricato, 2010).

According to Nobre et al. (2010) the RMSP could suffer a temperature increase of around 2 to 3°C this century and this level of climate change could cause a significant change in the rainfall regime. These authors predict that this situation would double the number of days with intense rainfall, with consequent flooding that would affect a larger area. Global warming will cause increasingly intense rainfall, so detention reservoirs (pools), which are currently overflowing with rubbish, will suffer serious damage if they are not designed in such a way as to make it difficult for waste and sediment to enter them (Giroldo, 2003; Besen et al. 2010).

1.2 MITIGATION MEASURES AND CONSEQUENCES

Since the 1960s, actions aimed at flood control have been developed in the RMSP. These initiatives gained more strength in the 1980s through various studies in which flooding was seen as an issue to be dealt with intersectorally (Jacobi, 2013).

In 1998, the Macrodrainage Master Plan for the Alto Tietê river basin was created, with the aim of combating flooding in the RMSP by building swimming pools, lowering the Tietê river channel, channelling tributaries and building dams (Jacobi, 2013).

Flood control or containment reservoirs, popularly referred to as "pools", change the way rainwater management is designed by trying to slow down the flow of liquid mass during episodes of heavy rainfall.

The 45 pools in the RMSP, 24 of which are the responsibility of the state and 21 of which are managed by municipalities, whether surface or underground, fulfil one of the functions of floodplains

- dampening the peak of floods - by releasing the water in a controlled manner and little by little after the end of heavy rain. Unlike natural floodplains, pools cannot be left unmaintained, as they quickly become breeding grounds for mosquitoes (Giroldo, 2003; Silvério, 2008; Rutkowski et al. 2010). Therefore, in order to guarantee the well-being and health of the surrounding population, it is necessary to monitor and control the mosquito populations associated with the pools.

Morrison et al. (2008) assess that analysing the density or frequency of adult mosquitoes is the most appropriate method for predicting the level of infestation or the level of risk of an epidemic emerging. However, they point out that monitoring is rarely implemented, perhaps due to the difficulty of operationalising it and the way it is collected to build indicators. This situation justifies the aim of this work, which will be to develop a monitoring strategy that is sensitive to detecting fluctuations in infestation levels in swimming pool areas, making it possible to issue warnings about health risk situations.

1.3 THE MOSQUITO Culex quinquefasciatus

The mosquito Culex quinquefasciatus, popularly known as the urban mosquito, muriçoca, carapanã, depending on the region of the country, is the target species of this study. This insect, common in urban areas with surface water sanitation problems, is classified in the order Diptera, family Culicidae (Forattini, 1962). The Culicidae family is large, with more than 3,500 known species currently distributed throughout the world, with the exception of the Antarctic Continent, a permanently frozen region (Forattini, 2002). Among the Culicidae, the species of interest is typically urban, with a high degree of synanthropy and anthropophily (Forattini, 2002).

The Culex quinquefasciatus mosquito, common in urban areas, is cosmopolitan. It occurs throughout Brazil, but its distribution and abundance are strongly influenced by the presence of humans, and it is found in human settlements and rural villages. It feeds on human blood at night, usually when people are resting (Consoli and Lourenço de Oliveira, 1994).

This culicid was known for a long time as Culex fatigans or Culex pipiens fatigans (De Meillon et al. 1967) in temperate areas, but the tropical variant was recognised as Culex quinquefasciatus. Their breeding sites can be artificial deposits, with water rich in organic matter, dirty and smelly in appearance. However, they usually occur in soil breeding sites, such as ponds, lakes and ditches polluted with human sewage waste. They can occur both in sunny breeding sites and in those protected by shade, always in the vicinity of human habitation. It is more frequent in the warm, rainy months, but is also collected throughout the year (Consoli and Lourenço de Oliveira, 1994).

According to Gomes and Forattini (1990), the adult forms of mosquitoes generally take shelter in the vicinity of their breeding sites or close to blood feeding sources; the most common shelter for Culex quinquefasciatus is human habitation or its surroundings.

Culex quinquefasciatus is reported to be highly anthropophilic, but in some publications, such as Laporta (2007), a greater philiasis towards hosts such as dogs and chickens was verified. In Forattini's review, the author refers to the possibility of this species being ornithophilic or saurophilic (Forattini, 2002). Thus, it is a species that can feed on a wide range of animals; however, its anthropophily is unquestionable in the literature.

Like most culicidae, the females of Culex quinquefasciatus are haematophagous and show marked endophilia and endophagy, i.e. they have the biological habit of preferring internal environments as well as feeding in these environments where they exert their blood repast, being practically the species that most bites humans inside houses at night; however, alternatively the females resort to complementary feeding on carbohydrates of plant origin (Consoli and Lourenço de Oliveira, 1994). As with other mosquitoes, the male feeds on carbohydrates found in plant sap, flowers and fruit and is not haematophagous. The female needs blood from hosts and her ovarian development depends on blood repast (Consoli and Lourenço de Oliveira, 1994). This phenomenon is called anautogeny. Those species that do not require blood to oviposit are called autogenous (Edman, 1997).

Culex quinquefasciatus is a nocturnal mosquito, so its haematophagous activity begins at dusk and reaches its maximum values after 10pm, with the highest frequency peaking around midnight. Pregnant females are found in the vicinity of breeding sites when they are about to oviposit, while newly engorged females are found in homes (Natal et al. 1991). This characteristic of resting after feeding is also found in males, which despite not feeding on blood have a marked endophilia. They accumulate in shady places inside houses under furniture and household utensils and are found in an equivalent proportion to females (Forattini, 2002).

The population density of the target species is variable and subject to climatic conditions, mainly rainfall. The duration in days of the adult mosquito, measured by the survival rate, is also dependent on the climate. The flight radius, usually estimated by the dispersal of adults from a point, can reach more than a thousand metres, but usually tends to be smaller when the food source and shelter are nearby (Forattini, 2002, Laporta, 2007).

When studying mosquito populations under natural conditions, Urbinatti (2000) cites biotic and abiotic factors as the variables that influence mosquito proliferation. Biotic factors include competition, parasitism and predation. Abiotic factors include: temperature, transparency, colour, pH, chemical composition of the water, suspended solids, Chemical Oxygen Demand (COD), Biological Oxygen Demand (BOD), dissolved oxygen in the water and electrical conductivity. Water cycles such as precipitation and evaporation directly influence both biotic and abiotic factors. The mosquito targeted in this study, Culex quinquefasciatus in its immature stage, is a species adapted to polluted waters, with a high organic load, high turbidity, low transparency, high nitrogen concentration, and can live in a practically anoxic environment, where its potential predators cannot survive. With plenty of food and no regulation of the ecological chain of predation, the species tends to proliferate in high density, becoming an urban pest.

Murrel & Juliano (2013) carried out a test in controlled natural shelters, which were tree hollows, separating groups of larvae into two groups. The first group contained larvae of Culex quinquefasciatus, Aedes aegypti and a predatory species of both, Toxorrinchites rutilus. In the second group there were only Culex quinquefasciatus and Toxorrinchites rutilus, and the researchers removed the Aedes aegypti eggs that were naturally laid by the free females. In both groups there was a decrease in Culex and Aedes, but in the first group there was a greater decrease in Culex quinquefasciatus larvae and pupae, demonstrating that the competition factor in this case was more important than predation as a factor restricting proliferation.

1.4 ZOONOSES AND OTHER DISEASES ATTRIBUTED TO CULICIDAE

Among the properties of mosquito vectors are vector competence and vector capacity. While competence deals with internal aspects of the organism in terms of receptivity to a given pathogen, to the point where the vector becomes infected and infectious, capacity is a broader issue, as it relates to the ability that the vector has under natural conditions to effectively fulfil its role as a transmitter. The Culex quinquefasciatus mosquito has vector competence and capacity for certain arboviruses and also for Wuchereria bancrofti (Forattini, 2002; Laporta, JL, 2004; Laporta, 2007).

Mosquitoes generally transmit a number of pathogens. Zoonoses associated with culicidae include: Eastern Equine Encephalitis, Western Equine Encephalitis, Venezuelan Equine Encephalitis (Calisher, 1982, Figueiredo, 2007), Rocio Encephalitis, San Louis Encephalitis, Ilheus Fever (Menezes, 2002), Yellow Fever, Oropouche Fever, Mayaro Fever, Guamá virus, Group A Alphavirus (Natal, 1981) and filariasis caused by nematodes (Acha and Szyfres, 1989). In Brazil, the mosquito that is the subject of this study is epidemiologically important because it is the proven vector of Wuchereria bancrofti, the agent of bancroftian filariasis, a disease that occurs in urban areas in the north-east of Brazil, and also because it is a potential vector of a series of arboviruses.

In Greater São Paulo, the main problem caused by the Culex quinquefasciatus mosquito is the nocturnal irritation it causes to humans when it bites. The reaction to this aggression varies from person to person, and there may be individuals with a high degree of sensitivity. There are individuals who are highly susceptible to the inflammatory response and who may suffer allergies and secondary bacterial infections of varying intensity, as well as blood contamination.

Notwithstanding the above, the common mosquito Culex quinquefasciatus, like other culicidae, has a characteristic buzzing sound and due to its nocturnal habits causes sleepless nights or poor sleep resulting in fatigue which can be the cause of accidents at work and in traffic (Kirby et al. 2008, Fofana et al. 2010, Mikulak et al. 2012, Biswas et al. 2013). Karch et al. (1993) in a study carried out in what is now the Democratic Republic of Congo found that in some urban areas of the capital Kinshasa the average number of bites per individual per night reaches 400 bites by the Culex quinquefasciatus species.

1.5 CONTROL MEASURES WITH SPECIAL REFERENCE TO Culex quinquefasciatus

With increasing concern for the environment and the advent of resistance that insects could present, various forms of control have been used, such as physical, chemical and biological methods.

The biological methods used for mosquitoes in general are highlighted: biological control through predation by arthropods such as crustaceans (copepods), insects (dragonflies and Toxorrinchites larvae), fish (particularly the Gambusia genus), parasitism through viruses, protozoa, fungi and helminths, and the use of biological larvicides that have bacterial toxins such as Bacillus thunrigiensis israelensis (BTI) and Bacillus sphaericus. As well as these, an organic and natural alternative is the use of plant toxins instead of synthetic compounds. These include derivatives such as: rotenone (obtained from the timbó vine), neem oil (whose active ingredient is azadirachtin), agave, garlic (whose active ingredients are a disulfide and a trisulfide) and pyrethrum (obtained from chrysanthemums or cloves) (Amonkar & Banerj, 1971; Margalit and Dean, 1985; Grainge and Ahmed, 1988; Jones et al. 1990; Perich et al. 1990; Chadad & Boff 1994; Santos et al. 1994; Rodriguez, 1995, Pizarro et al. 1999, Cuba, 2004). For the target species of this study, as its immatures are found in highly polluted aquatic collections, direct predation is unfeasible, as most predators do not live in environments without oxygen. However, bacterial toxins have been used successfully in mosquito control programmes (Morais, 2005). A product originating from the fermentation of a soil bacterium (actinomycete), Sacchropolyspora spora spinosa has shown promise in larval control (Jiang & Mulla, 2009). Rasgon (2012) sees good prospects for the use of bacteria of the Wolbachia genus in the control of culicidae.

The use of growth inhibitors (GRIs) such as triflumuron and diflubenzuron, although more expensive, should be emphasised. These products act on the formation of the chitin protein that makes up the exoskeleton of arthropods, so they are targeted at that animal phylum, but the problem is the low persistence of these compounds in the environment, as well as their cost (Victor Neto, 2006). It is likely that this class of insecticide is not suitable for the polluted environments where Culex quinquefasciatus proliferates. Furthermore, despite the high specificity of the physiological action of growth inhibitors, resistance to these compounds has been observed in houseflies, in addition to the well-known and well-studied resistance of insects to organochlorine, organophosphate and pyrethroid insecticides (Cerf & Georghiou, 1974; WHO, 1997; Zhang et al. 1998, Bracco, 1998, Vitor Neto, 2006, Portal da Saúde, 2013).

1.6 MONITORING THROUGH ENVIRONMENTAL MANAGEMENT INDICATORS

Because of the phenomena described, it is essential to mitigate possible environmental and biological damage from the application of any control products in the environment. In this context, the issue of monitoring becomes more relevant, with the aim of using control at the right times and in the right doses, whether with spatial or focal application strategies. One of the challenges facing Public Health is to structure indicator systems that make it possible to analyse current conditions and trends in interactions between driving forces and socio-economic pressures and progressive environmental degradation, affecting the services of ecosystems that support human life, health and well-being. With this in mind, recent efforts have been made in the country to create indicators of environmental sustainability and health and well-being on a municipal and micro-regional scale for municipalities that have had a large part of the Atlantic Forest biome destroyed, mainly since the 18th century (Freitas & Giatti, 2009).

An index is the numerical expression of indicators, and can also be made up of a composite of data or other indicators that make it possible to monitor the interactions between the environmental and socio-economic components, as well as a database for strategic planning and policy analysis to identify sustainable development pathways. In this way, it helps governments to formulate and monitor economic policies more effectively, to determine strategic environmental regulations for more effective resource management and to use taxes and subsidies more efficiently. It also offers a way of improving political dialogue between different actors by providing a transparent system of information on the relationship between human and environmental activities (Cezare et al. 2007, Philippi Jr & Malheiros, 2012).

Although the Municipality of Santo André has a robust environmental policy with the creation of the Environmental Management System, based on law, the creation of sustainability indicators that will affect the health area from a multidisciplinary perspective is still in its infancy (Cezare et al. 2007).

In this conception of sustainability, health cannot be treated as dissociated from well-being and socio-environmental changes and their impacts on ecosystems, so that even mediated effects on health, measurable today, can result from multiple interactions at the end of a long and complex network of events and situations combined on different spatial and temporal scales, ranging from environmental changes that alter the distribution and behaviour of vectors and hosts, resulting in epidemics of dengue, malaria and leishmaniasis, or even events such as droughts and floods that will exacerbate infections or infestations originating in drinking water or wastewater. In addition to these effects, we must also consider a long-term perspective for sustainability, in which broad socio-environmental changes can result in emergent properties with radical alterations or disruptions in ecosystem services and major effects on health, requiring long-term recovery or adaptation (Freitas & Giatti, 2009).

According to Domingues (2000) apud Philippi Jr & Malheiros (2012), with the growing demand for information that portrays environmental problems and the concern about medical and health issues, especially in urban areas, there has been an attempt to identify interfaces between problems that originate in different sectors, in order to develop new types of information that explicitly demonstrate the relationship between economic development and unsatisfactory health conditions due to environmental degradation.

The environment is undoubtedly an important component of the system under discussion and annual variations in temperature can have an influence on various diseases. Practically all those diseases that are transmitted by vectors are favoured in the warmer periods of the year. Insects such as mosquitoes and flies proliferate more intensely at higher temperatures when their densities increase. Pathogens are then transmitted more easily due to the greater frequency of contact between these vectors and humans or their food (Natal et al. in Philippi Jr & Pelicioni, 2005).

There is a notorious need to adopt mosquito and vector monitoring indexes in the professional environment, in order to guide actions to control and combat the different problems that affect the social environment. This system is common in official state circles, making it an important source of information. On the other hand, there is pressure from the media and society in general, including academia, because it is by monitoring these indices or indicators that the population can exercise control as part of the free exercise of citizenship.

In the field of Medical Entomology, for example, we have the entomological indicators of the National Dengue Control Plan. These indicators have already been adopted by more than 4,000 Brazilian municipalities. One example is the Aedes aegypti Rapid Index Survey (LIRAa). This is a type of larval infestation analysis that is recommended to be carried out every October. One cannot fail to recognise the Ministry of Health's efforts to standardise data in a country of continental dimensions, with such disparate regions in terms of climate, terrain, socio-economic factors, cultural, environmental and demographic diversity. The entomological indicators obtained in the LIRAa are the Breteau index (BI), which shows the ratio of positive breeding sites to the number of properties surveyed, the Building index (PI), which shows the number of positive properties to the number of properties surveyed and the Container index (RI), which is the ratio of positive containers to the number of containers surveyed. This last index should be analysed together with the two previous indices, but it has an inverse relationship with them, predicting possible future infestation (MS, 2005). The decentralisation of vector control actions, which is in the process of being implemented in Brazil, has led to the need to incorporate into the health responsibilities of states and municipalities activities carried out at people's homes and workplaces, activities that were not part of the traditional collection of the vast majority of states and municipalities, as medical care activities already were. Recruiting staff, training them, supervising and monitoring them to carry out complex field activities, such as identifying anti-vector measures suited to the different local realities and applying them, has been beyond the capacity of the vast majority of municipalities, especially small ones. Added to this is the fact that budgetary and financial constraints have made it difficult to create reasonably paid professional careers to keep contracted staff, usually with precarious labour relations, on the job long enough to acquire fieldwork experience. These aspects have posed a major challenge for the country's public health system. Different control models have been considered and many strategies have been questioned and modified (Tauil, 2006). As a result, in the practice of the programmes, there are discrepancies between what these indices indicate and reality. These evaluations are subject to human errors such as miscalculations and the configuration of the service itself, such as residents refusing to allow the agent to enter their homes, especially in critical areas. However, these facts are exceptions and in general the application of the index achieves its objectives, which is the numerical interpretation of dengue mosquito infestation in a vast area of the national territory (MS, 2005).

Methods for collecting adult mosquitoes in order to build indicators for Aedes aegypti, such as human bait (Okumu, 2010), such as aspirators or traps, have not been routinely used by the Ministry of Health (MS, 2011). This position is due to the risk of exposure of field operatives to infected mosquitoes and also because the time for evaluating specificity and efficiency is very limited. Monitoring the population of adult mosquitoes would have advantages such as providing a better estimate of the density of the vector at the stage of life in which it fulfils its role and more directly ascertaining the risk of transmission. Also, collecting winged mosquitoes allows for a better analysis of spatial distribution, parity, ovarian development, intestinal content to verify repassing, susceptibility to pathogens and resistance to insecticides (Barata et al. 2001).

David et al. (2009) observed that Aedes aegypti females marked with phosphorescent powder and released and recaptured have a lower survival rate in upper middle class neighbourhoods than in suburbs and slums, as do infestation rates measured by trapping and aspiration. In view of their data, the authors proposed that there should be a better distribution of resources in a more cost-effective way due to the urban structure that should be evaluated taking into account these biological parameters.

1.7 PROBLEMATISATION OF THE RESEARCH

Unlike the control of Aedes aegypti, the dengue virus vector, which has a series of practical and useful entomological indicators to support surveillance, there are no properly standardised monitoring indices for Culex quinquefasciatus to guide its control in the areas impacted by the pools. This study focuses on this point, as it is well known how difficult it is to keep these reservoirs clean and well maintained in order to avoid the proliferation of mosquitoes (Culicidae). Developing a practical and effective way of monitoring the population of these dipterans suitable for these areas will make control intervention strategies easier and more rational. A study aimed at creating and standardising such indices for monitoring this urban pest is therefore justified.

CHAPTER 2

OBJECTIVES

2.1 GENERAL OBJECTIVE

To establish Culex quinquefasciatus infestation rates, based on the count of adult insects and their immature forms collected in swimming pools, and to propose a tool for estimating the nuisance caused by bites in the surrounding population.

2.2 SPECIFIC OBJECTIVES

Construct and validate monitoring indices for immature and adult mosquitoes.

To validate the method of collecting adult mosquitoes by aspiration in shelters and resting boxes and to validate the collection of immatures for entomological monitoring purposes in swimming pool areas by analysing the proposed indices.

To verify aspects of the seasonality of Culex quinquefasciatus proliferation in swimming pools and determine the best time for control interventions.

Obtain and validate the population's annoyance indices in relation to mosquito bites in the surroundings of swimming pools.

CHAPTER 3

BRIEF PRESENTATION OF THE METHODS ADOPTED

Initially, a literature review was carried out using the PUBMED database and, to a lesser extent, LILACS. Descriptors such as mosquito control, swimming pools, flood containment reservoirs, nuisance, Culex were used.

In addition to the articles found, the review was also carried out by consulting articles, books and periodicals in the Public Health Library and the Civil Engineering Library of the Polytechnic School (POLI-USP). Books from personal collections were also consulted, as well as data, articles and books from the entomology laboratory and from the supervisor himself.

Because the author's professional practice is in an official body, manuals, projects and official legislation were used. The courses taken in the Professional Master's in Environment, Health and Sustainability provided literature mainly in the environmental area, with emphasis on Environmental Management Systems, Interfaces in Environment, Health and Sustainability, Environmental Health Seminars, Hydrobiology of Water Supply and Sewage Treatment and even Ergonomics were important in providing notions and concepts for the evaluation used in the Discussion.

After acquiring the material to be used in the field, and in some cases preparing the material, such as the resting boxes which were painted on the inside as well as making an inlet hole, the practical work of sampling the two flood containment reservoirs began.

A checklist was drawn up with the material and procedures for each activity to be carried out, of which there were three: Collecting Immatures, Vacuuming and Vacuuming Resting Boxes.

Samples were collected from August 2014 to May 2015, once a month in each reservoir. After formal interviews with active and retired GCZ employees, residents and employees of companies in the region of the pools and the watchmen of the RM-9 Faculdade de Medicina Reservoir (which is the ABC Foundation Medical School), the sampling points for the collection of immature forms were chosen: five for each reservoir, five suction points for each reservoir and five points for the installation of the resting boxes in each reservoir.

The immature collection points were named LV (LV1, LV2, LV3, LV4 and LV5) for the points in the Valparaíso pool and LBP (LBP1, LBP2, LBP3, LBP4 and LBP5) for the points in the Bom Pastor pool.

The points where the adult forms were sucked into the vegetation were named AV (AV1, AV2, AV3, AV4 and AV5) for the points in the Valparaíso pool and ABP (ABP1, ABP2, ABP3, ABP4 and ABP5) for the points in the Bom Pastor pool.

The resting boxes were named CR and were installed in the Valparaíso pool (CR1, CR2, CR3, CR4 and CR5) and the Bom Pastor pool (CR6, CR7, CR8, CR9 and CR10).

All the sampling points for the three activities were the same from the beginning to the end of the fieldwork.

The first collections were initially made in the two reservoirs on a single day, but the reservoirs were visited the previous day during the twilight period for the installation of the resting box. The following day, the work consisted of closing and collecting five resting boxes per reservoir, collecting immatures using the entomological shell at the five determined points and then aspirating them with the Nasci aspirator at five determined points. The aspirated mosquitoes were killed with chloroform and placed in entomological boxes. After the team arrived at the laboratory, the resting boxes were carefully opened to insert the Natal and Marucci aspirator and collect the mosquitoes, which were killed with chloroform and placed in separate entomological boxes. The counting of larvae and pupae and the identification and counting of adult mosquitoes was carried out in the GCZ entomology laboratory.

From October onwards, the work team was reduced and the work was divided into three days: the first day in the afternoon, during the twilight period, the boxes were installed in the Valparaíso pool; the next day, in the morning, samples of immatures were collected and adults aspirated, as well as the collection of CRs in the Valparaíso pool; in the afternoon, the boxes in the resting boxes were installed in the Bom Pastor pool and the Valparaíso pool was identified and counted in the laboratory. On the third day, in the morning, samples of immatures were collected and adults aspirated, as well as the collection of CRs in the Bom Pastor pool. In the afternoon, the Bom Pastor pool was counted and lab-identified.

After working in the laboratory, the data was tabulated in Excel and six spreadsheets were created, namely: (1) Monthly number of immatures per sampling point, (2) Monthly number of immatures per development stage, (3) Number of mosquitoes aspirated in the field, (4) Monthly number of adult Culicidae collected from resting boxes, (5) Number of females according to physiological state, (6) Graphs and Indexes spreadsheet.

These spreadsheets served as the basis for this dissertation and were supplemented with climatological and operational data from SEMASA (Santo André Municipal Environmental Sanitation Service) and DAEE (Department of Water and Electricity) and insect control data and complaints from residents at PMSA (Santo André City Hall).

CHAPTER 4

MATERIALS AND METHODS

4.1 STUDY AREA

The study areas defined for this research are located in the urban centre of the municipality of Santo André. They correspond to two swimming pools designed to regulate rainwater during flood peaks.

Figure 1 below shows the urban network of Greater São Paulo, indicating the city centre of Santo André.

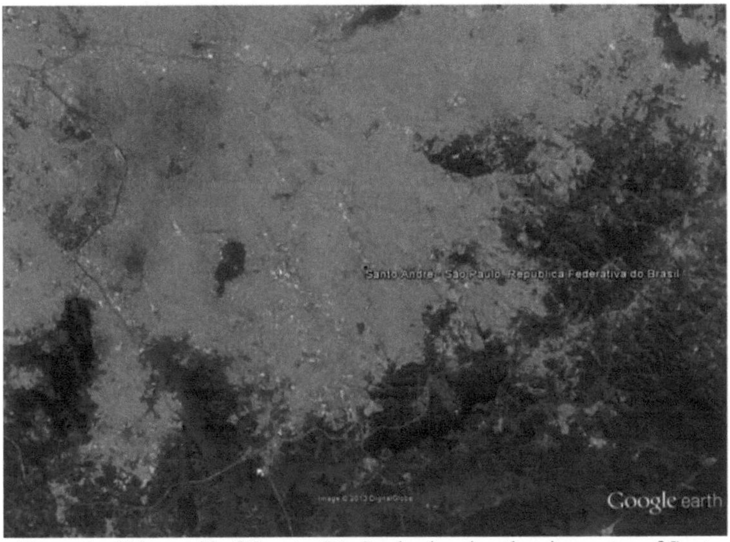

Figure 1 - Urban grid of Greater São Paulo showing the city centre of Santo André, SP. Source: Google earth.

Figure 2, shown below, shows part of the urban network of Greater São Paulo, containing the boundaries of the municipality of Santo André.

Figure 2 - Urban grid of Greater São Paulo showing the boundaries of the municipality of Santo André, SP, highlighted in red. Source: Google earth.

The municipality of Santo André has five swimming pools, three of which are municipal and two state-owned. Although the Department of Water and Electricity (DAEE) is responsible for cleaning the equipment, the Santo André Municipal Environmental Sanitation Service (SEMASA) weeds the reservoirs an average of six times a year, as well as carrying out mechanised cleaning twice a year. The state reservoirs are RM 9 (Faculdade de Medicina) and RO 4 (Oratório). The municipal units are Santa Terezinha, Bom Pastor and Vila América. For this project, two reservoirs in the municipality of Santo André were chosen: the state reservoir located on Rua Grã-Bretanha (Faculdade de Medicina) and the municipal reservoir located on Rua Xingu (Bom Pastor), as they have a history of the most common mosquito complaints from residents of the city.

3.1.1 Medical School Reservoir (Valparaiso)

One of the areas chosen for research is the so-called RM-9 Reservoir/Medical School. This pool is located in Rua Grã-Bretanha, Vila Príncipe de Gales. Although the official name of the neighbourhood is Vila Príncipe de Gales, due to the avenue of the same name (which is relatively far away), this reservoir is popularly known as the Valparaíso reservoir because it is in this neighbourhood, where the Zoonosis Control Department is also located, less than 200 metres from the reservoir. It has been operating since 1999 and has a capacity of 120,000 metres^3 . It is classified as "off-line". **Figure 3**, shown below, shows an aerial image of this swimming pool. **Figure 4** shows a photograph of the study area.

Figure 3 - Aerial image of the RM-9/Faculdade de Medicina reservoir, located in the urban area of the municipality of Santo André, SP. Source: www.semasa.sp.gov.br

Figure 4 - View of the RM-9/ Faculdade de Medicina reservoir, located in the urban area of the municipality of Santo André, SP. Photo: Delsio Natal

The surroundings of this reservoir are characterised by being in a typically urban area, next to the Ribeirão dos Meninos, which is a stream used to delimit the political-administrative boundary of the municipalities of Santo André and São Bernardo do Campo.

In its immediate surroundings we will find administrative buildings such as the ABC Foundation Medical School on the north side and the Association of Municipal Civil Servants and the Department of Parks and Green Areas on the south side, these being public bodies with large areas of vegetation and recreation, in other words, with buildings but no homes.

If you look at the east side, you'll find homes and services such as mechanical workshops less than ten metres from the edge of the swimming pool. The economically active population is made up of middle-class and lower-middle-class people living in brick houses.

On the west side we will have Avenida Lauro Gomes, an expressway that borders the border between the two municipalities and the Ribeirão dos Meninos.

3.1.1.1 Sampling Points for the Collection of Immature Forms

All the points were chosen after an interview with the watchmen at the Valparaíso swimming pool, based on questions about where the most permanent water collections occurred. Once this was done, the following points were chosen and named for larval collection (photos of the points are in Appendix A):

Point LV1 - Near the spillway.

This is the water inlet to the Swimming Pool and is located on the south-west side. The water is usually dirtier at this point and is only filtered initially.

Point LV2 - Near the pumps

It is located in front of the pumps that will discharge the incoming water. It is usually a little cleaner than the previous point, as it has undergone natural settling processes. It is a point that can change the collection point if the Pool is completely flooded because it is on the west side right in the middle.

Point LV 3 - Northwest Pool

It is located on the north-western side and is a nesting site for quer-queros and irerês. It is difficult to access, with the risk of the sample collector getting bogged down, as well as the birds flying overhead.

Point LV 4 - Edge of the bush

It is located on the south side of the pool and although it collects water, it is perennial and subject to desiccation. When the pool is clean, we can see that it is a channel.

LV 5 - Channel of the great wall

It is located on the eastern side of the pool and is perennial, suffering from desiccation in the drier months.

Aerial image of the sampling points for larvae collection in the RM-9 Medical School Reservoir

(Valparaiso), labelled LV followed by a number and highlighted in yellow. Source: Google Maps.

4.1.1.2 Aspiration Sampling Points for Adult Forms

The suction points in the Valparaíso swimming pool were chosen because they were in a well-defined and delimited area, i.e. they were chosen to be equidistant, so one point was chosen on the south-east side, one on the south-west side, one in the centre, one on the north-west side and one on the north-east side. The point chosen in the centre of the swimming pool could possibly be replaced by a point on the north face, in the middle of the side or on the north-east and north-west corners when the swimming pool is flooded, making suction impossible, but this situation did not arise during the fieldwork. The suction points in this swimming pool have been labelled with the acronym AV and its corresponding number (photos of the points can be found in Appendix A).

Point AV-1 - Input

As its name suggests, it is located at the entrance to the pool near the watchmen's quarters and is on the south-east side.

Point AV- 2 - Ramp

It is located on the south and south-east sides and is characterised by a cobbled ramp for the entrance of employees and machinery.

Point AV-3 - Midfield

It is characterised by its location in the centre of the pool, in a place prone to flooding.

Point AV-4 - Corner of Avenida

It is located on the north-west side and this point is characterised by the accumulation of rubbish, especially pet bottles. Behind the wall is Avenida Lauro Gomes, which borders the border between the municipality of Santo André and the municipality of São Bernardo do Campo.

Point AV- 5 - Grandstand.

It is located on the north-east side of the pool and is so called because it has a court with bleachers, a recurring idea among designers to turn pools into leisure areas.

Figura 6. Aerial image of the aspiration sampling points in the shelters for collecting adult mosquitoes in the RM-9 Medical School Reservoir (Valparaiso), labelled AV followed by a number and highlighted in green. Source: Google Maps.

4.1.1.3 Restroom Installation Points

The installation of the resting boxes was based on a mixture of reports from the pool's watchmen and geographical location to maintain a more or less regular distance between the boxes. The boxes were named CR and are numbered from 1 to 5 and were installed in the Valparaíso pool. (Photos of the installed boxes are in Appendix A).

CR 1 - Watchmen's accommodation

It is located on the south-east side and was installed because of reports from the watchmen of a night-time nuisance caused by mosquitoes.

CR 2 - Amoreira

This point was chosen because mosquitoes are known to prefer to shelter under trees due to their biological characteristics. As there are few trees in the pool area, this point corroborates this situation. It is located on the south side.

CR 3 - Midfield

Installed in a place where it can be replaced in the event of flooding. It is located in the centre of the pool.

CR 4 - Near the pumps

It is located on the west side of the pool, in an area prone to flooding.

CR 5 - Staircase

It is located on the north-east side of the pool next to a concrete staircase.

Figura 7. Aerial image of the sampling points of the resting boxes installed to collect adult mosquitoes in the RM-9 Faculty of Medicine Reservoir (Valparaiso), labelled CR followed by a number (1 to 5) and highlighted in blue. Source: Google Maps.

RM-9 Reservoir Faculty of Medicine (Valparaiso)

Immature collection (LV)	• LV 1 Spillway • LV 2 Pumps • LV 3 Northwest Pool • LV 4 Borda do Mato (Canaleta) • LV 5 Great Wall Canal
Vacuum in shelters (AV)	• AV 1 Input • AV 2 Ramp • AV 3 Midfield • AV 4 Corner Avenue • AV 5 Grandstand
Vacuuming Restrooms	• CR 1 Watchmen's quarters • CR 2 Amoreira • CR 3 Midfield • CR 4 Near the pumps • CR 5 Staircase

Figura 8. Sampling points chosen for the samples taken from August 2014 to May 2015 at the RM-9 Faculty of Medicine Reservoir (Valparaiso).

4.1.2 Bom Pastor Reservoir

The other area chosen for research is located in the Várzea do Ribeirão dos Meninos, built next to the bed of the Araçatuba stream, from Rua Dom Silvério Pimenta to Rua Xingu. This reservoir has been in operation since 1994 and has a capacity to retain 19,300m^3 of water, making it classified as "on-line". On-line reservoirs are those whose construction follows the bed of the river or watercourse, while off-line reservoirs are on the bank of the river or watercourse (Giroldo, 2003, Silvério, 2008). The image in **Figure 9** shows the urban fabric of Santo André, highlighting the swimming pool, which has a linear configuration, following the bed of its host stream. **Figure 10** shows a photograph of the same pool from a ground-level perspective.

Figura 9. Satellite image of the Jardim Bom Pastor Retention Reservoir, located in the urban area of the municipality of Santo André, SP. Source: Google earth.

Figure 10. View of the Jardim Bom Pastor Retention Reservoir, located in the urban area of the municipality of Santo André, SP. Photo taken on 11 September 2013. Photo: Delsio Natal.

The surroundings of this reservoir are also characterised by being in a typically urban area, flowing into the Ribeirão dos Meninos at its northern limit.

This municipal swimming pool is characterised by being in an open area, without well-defined boundaries by fences and walls, and we can divide it into two parts to explain it better. The first part consists of a circular lagoon flanked by transport and car repair companies, as well as a landscaping company. Access to this lagoon is via Rua Xingu, in the Bom Pastor neighbourhood. There are brick houses belonging to people with low economic power around the lagoon. The second part consists of an arm of water that leaves the lagoon through a space made for electricity transmission line towers and runs through the residential neighbourhood of Vila Floresta, passing under paved streets and wooden pinguelas built for pedestrians.

This arm of water, due to the action of aquatic plants that grow there, sometimes has extremely clean water at certain points and we even find frogs, which is an indicator of good water quality despite being in a strictly urban environment. It should be emphasised that this situation is not constant in both space and time, as there are times and stretches of this stretch of the pool where there is a high proliferation of mosquitoes that thrive in polluted water, and because this second part is entirely flanked by residential and commercial streets with masonry houses, it provokes complaints from residents who live or work there.

The pool's area was reduced due to the intervention of the government, which built a low-income housing estate on its southern boundary where there used to be a housing estate. As a result of the building work, the backwater of the Araçatuba stream, which was part of the Bom Pastor pool, was filled in, separating the running waters of the stream from the pool.

Due to this discontinuity between the stream and the pool, the new boundaries of the pool suffer greater desiccation during dry periods, further reducing its area.

9.1.2.1 Sampling Points for the Collection of Immature Forms

The choice of collection points for larvae and pupae in the Bom Pastor reservoir was based on the experience of some of the Zoonosis Control Management staff and also on the intention of distributing the points along the length of the reservoir, which stretches for several kilometres. They were labelled LBP followed by the number (photos of the points are in Appendix A).

Point LBP 1 - Rua Xingu, the pool itself, is characterised by its circular shape.

LBP Point 2 - Rua Xingu, near the horses

There is clandestine breeding of horses in this municipal pool, and the owner of the animals has built a barbed wire fence in the area of the pool (see figure 30 in Appendix A) and this site was chosen as a reference for larval collection.

LBP Point 3 - Rua Felício Pedroso

The pool, as described above, has a "wetland" character and borders Felício Pedroso Street, where point LBP 3 has been designated. In the background of the image shown in figure 31 of Appendix A, we can see the homes from which residents have complained about mosquitoes. Wetlands are artificially flooded lands, which differentiates them from marshes where flooding is natural and the purpose is to purify the water through aquatic plants, feed animals or for urban beautification and to improve environmental conditions (Medland & Vaux, 2014).

LBP Point 4 - Manilha (Rua Inverno com Dom Silvério Pimenta)

This point was chosen because of several complaints from residents of these streets.

LBP Point 5 - Rua Iguape

Despite being almost at the end of the pool's reach, it is a favourable place for immature forms to grow.

Figure 11. Aerial image of the sampling points for collecting larvae in the Bom Pastor Retention Reservoir, with its outline highlighted by the red line and whose points have been labelled LBP followed by a number and highlighted in yellow. Source: Google Maps.

4.1.2.2 Aspiration Sampling Points for Adult Forms

The choice of aspiration points in the Bom Pastor pool was based on interviews with residents and employees of commercial establishments close to the pool area. They were labelled ABP preceded by a corresponding number (photos of the points are in Appendix A).

ABP 1 - Rua Xingu (Near the urucum tree)

At this point, which is inside the pool itself, there is a wall that is discontinuous, with the eastern part of the wall having the reference of an annatto tree. Along this wall there are reports of high infestations of adult mosquitoes, a fact that was proven when the collections took place when the undergrowth was high.

Point ABP 2 - Near the sign.

The aspirated area is close to the wall, on its western side, where a billboard was used as a reference.

Point ABP 3 - Near the Winter Street footbridge.

This point was chosen due to complaints from residents of this street

Point ABP 4 - Near the Osasco Street footbridge

This is another pinguela taken as a reference.

Point ABP 5 - Rua Iguape

The site was chosen because it had a large number of immatures and should logically be a collection point for adults.

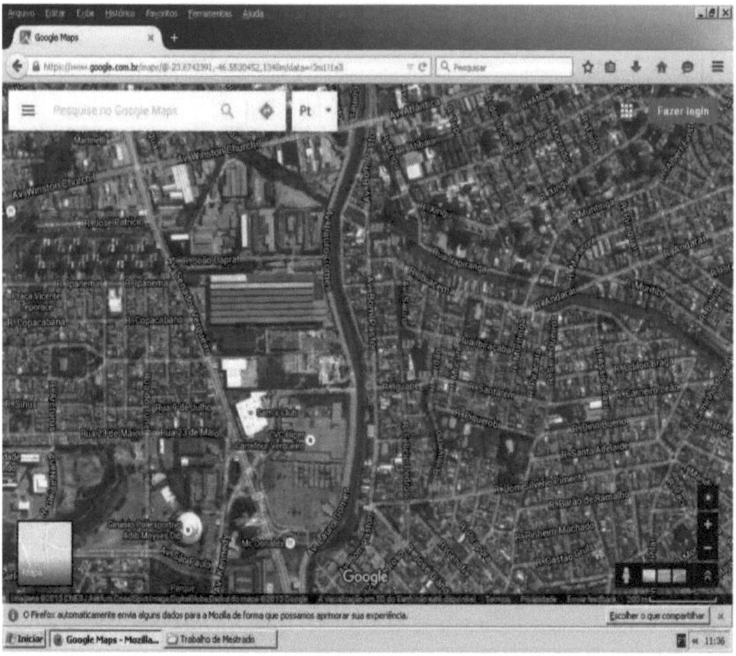

Figure 12. Aerial image of the adult mosquito aspiration sampling points in the Bom Pastor Retention Reservoir, with their outline highlighted by the red line and whose points have been labelled ABP followed by a number and highlighted in green. Source: Google Maps.

4.1.2.3 Restroom Installation Points

The resting boxes installed in the Bom Pastor pool were numbered from 6 to 10 and unlike those installed in the Valparaíso pool, the criteria used to choose them, in addition to their geographical location and reports of complaints, had to take into account another item, which was that they would not be disturbed by people or animals because they are in an open area and even with the concern to keep them hidden, they were sometimes tampered with and some boxes even stolen and had to be replaced with others. They were labelled with the acronym CR and the corresponding number (photos of the boxes can be found in Appendix A).

CR 6 - Cemented

It is located at the northern end of the pool where it drains into the Ribeirão dos Meninos. Extremely polluted water.

CR 7 - Shrub

It's opposite a gardening company, whose employee suggested the places where, according to him, there is the highest concentration of adult mosquitoes. It's on the west side of the pool itself (lagoon).

CR 8 - Rua Felício Pedroso, at number 48

It's a place that provides shade for mosquitoes and can be hidden from public view, as well as complaints from residents of this street.

CR 9 - Under the Winter Street footbridge

In addition to reports of complaints, it was installed to protect against the curiosity and manipulation of passers-by.

CR 10 - Under the Osasco Street footbridge

In addition to reports of complaints, it was also installed to protect against the curiosity and manipulation of passers-by.

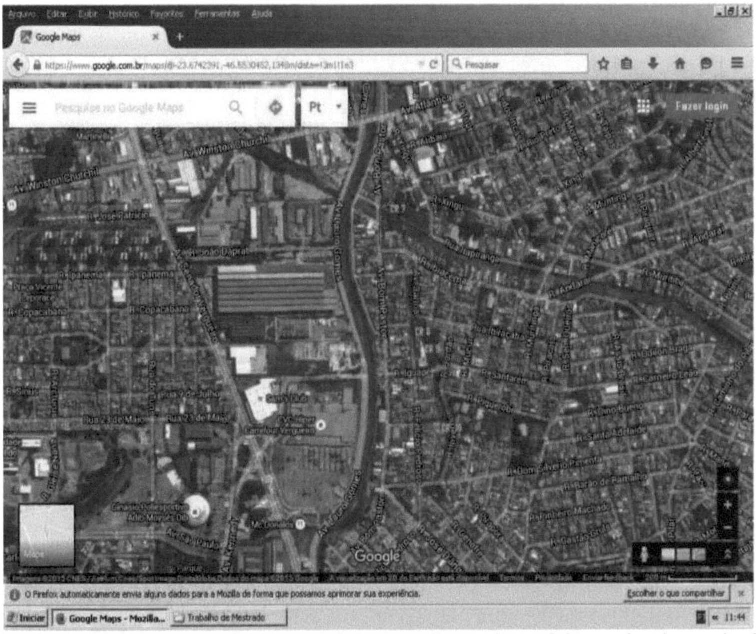

Figura 13. Aerial image of the sampling points of the resting boxes in the Bom Pastor Retention Reservoir, with their outline highlighted by the red line and whose points were named CR followed by a number (6 to 10) and highlighted in blue. Source: Google Maps

Bom Pastor Retention Reservoir

Collection of immatures (LBP)	• LBP 1 Lagoa • LBP 2 Near Horses • LBP 3 Rua Felício Pedroso • LBP 4 Rua Inverno (Rua Manilha) • LBP 5 Rua Iguape
Aspiration in Shelters (ABP)	• ABP 1 Annatto tree • ABP 2 Outdoor • ABP 3 Winter Street • ABP 4 From Osasco Street • ABP 5 Rua Iguape
Vacuuming Restrooms	• CR 6 Cemented • CR 7 Shrub • CR 8 Rua Felício Pedroso • CR 9 Under the Winter Street footbridge • CR 10 Under the Osasco Street footbridge

Figure 14. Sampling points chosen for samples taken from August 2014 to May 2015 at the Bom Pastor Retention Reservoir.

4.2 COLLECTION METHODS

4.2.1 Immature Forms

The immatures were sampled in the water collections that form in the spillway, channel or even in puddles that appear inside the pool or on its edges. The best collection points were chosen on the basis of observation criteria at the breeding sites and interviews with the local population and employees of nearby companies. The points selected for sampling remained the same until the fieldwork was completed.

For the specific collection of larvae and pupae for the construction of immature indices, a 350 ml entomological shell was used, to which an extension cable was attached. Field visits took place once a month throughout the observation period. At each visit, samples were obtained by immersing the shell in the water collection until it reached one litre (1000 ml), and the water collected was transferred to a graduated flask. Five points were defined in each pool, based on Ministry of Health standards (MS, 2011). First- and second-stage larvae were discarded because they were small and difficult to visualise.

Third and fourth stage larvae and pupae were counted. Identification and counting took place at the Entomology Laboratory of the Zoonosis Control Department of Santo André City Hall.

4.2.2 Adult forms

The adult mosquitoes were collected using two techniques: electric aspirators in the shelters and in the resting boxes.

4.2.2.1 Aspiration

The aspiration method for obtaining winged mosquitoes consists of using a manual aspirator with a "PVC 6" body, a 12-volt motor, an aluminium tube 60 cm long and a dipper. This equipment is powered by a battery which is attached to the collector's body by means of a leather strap (Nasci, 1981). Five equidistant points were sampled during each monthly field visit. These points are located around the flooded area, in places covered by vegetation, trees or grasses, which represent the shelters of adult mosquitoes. Each point was vacuumed, with a timed operation of five minutes. The adult mosquitoes were removed from inside the bucket and then killed with chloroform. They were placed in standardised entomological boxes and sent for identification and counting to the Entomology Laboratory of the Zoonosis Control Department of Santo André City Hall.

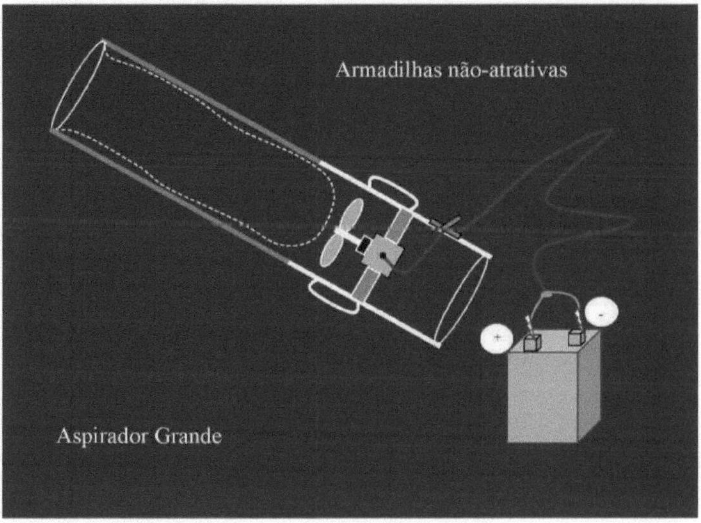

Figure 15 - Drawing of a Nasci Vacuum Cleaner. The dashed line corresponds to the filo where the adult forms of the mosquitoes are collected.
Source: Paulo Urbinatti.

4.2.2.2 Rest boxes

As this is a collection technique that is little used in Brazil, the resting box collection strategy adopted in this study took into account aspects described by various groups of researchers in different parts of the world (De Meillon et al, 1967; Komar et al., 1995; Edman et al., 1997; Harbison et al., 2006; Braks and Cardé, 2007; Barata et al., 2007; Kweka et al., 2009; Van den Bijllaardt, 2009; Govella et al., 2011; Manda et al., 2011).

Five resting boxes were set up for sampling adult mosquitoes in the areas of the pools selected for the study. Each of these units consisted of a Styrofoam box, 36 cm high, 40 cm long and 28.5 cm wide, painted on the inside with matt black latex paint. A circular opening was made in the centre of the removable lid for mosquito access, approximately 15 cm in diameter, so that the portable hoover and the operator's hand could be inserted. The lid was closed and its edges sealed with adhesive tape. The box was positioned so that its opening remained on the side, making it easier for mosquitoes to enter. The five boxes were positioned at equidistant points around the flooded area, in places favourable to mosquitoes. To make them more attractive to culicidae, in addition to the black paint inside, a damp cloth was placed inside each box to simulate an optimum humidity microclimate, which is essential for this type of insect. In each pool, the five boxes were installed once a month in the evening twilight around 5pm. The boxes were visited the following day at around 11am in the morning, when the adult specimens were collected from each box using a portable hoover. This device has a "PVC 3" body, a 6-volt micromotor, a plastic shaft and propeller and a steel mesh funnel collection cup (Natal & Marucci, 1984). In the time between collecting the box and aspirating it in the laboratory, the opening of the box was sealed with a sheet of A4-sized sulphite paper and adhesive tape around the edges of the paper. On arriving at the laboratory, the sulphite sheet was carefully removed from the opening of the box and the aspiration was carried out inside the box. The aspirator collection cup containing the mosquitoes was transferred to a plastic bag containing cotton wool soaked in chloroform. The dead mosquitoes were placed in labelled entomological boxes and sent for identification and counting to the Entomology Laboratory of the Zoonosis Control Department of Santo André City Hall.

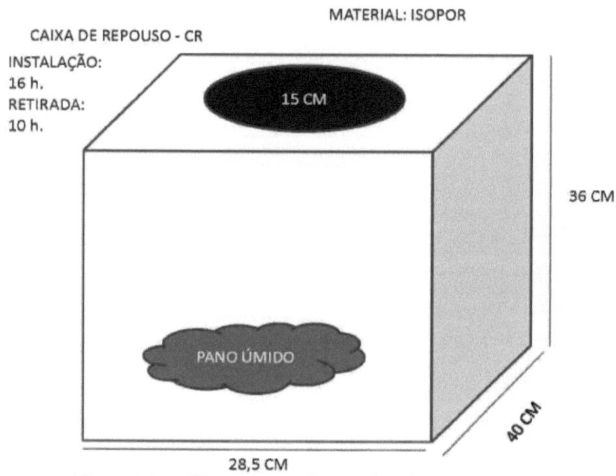

Figura 16. - Illustration of a resting box.

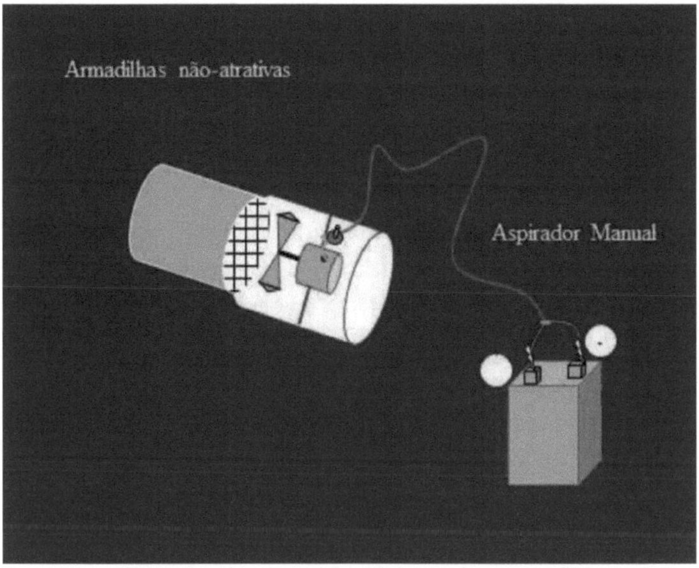

Figura 17. Illustrative drawing of the Christmas hoover and Marucci. Source: Paulo Urbinatti

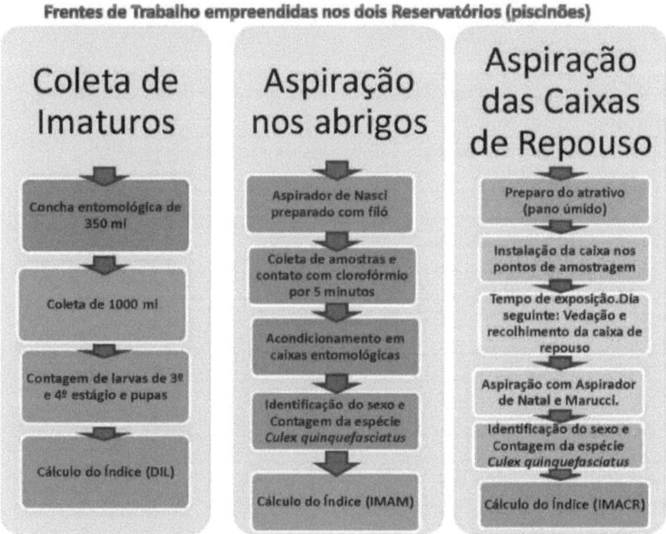

Figura 18. Flowchart of field activities carried out at the RM 9 Faculty of Medicine Reservoir and Bom Pastor Retention Reservoir, from August 2014 to May 2015.

4.3 NUISANCE ANALYSIS

The following strategy was adopted to build indicators to assess the nuisance caused by mosquitoes to the population around the pools.

4.3.1 Analysis by the Customer Service Information System

Using georeferencing applications, the area was delimited within a circular radius of 2000m from the geometric centre of each pool chosen for study. Within the limits of this coverage area, complaints about mosquito nuisance or requests for assistance from residents regarding infestation by the common mosquito were recorded. Every complaint detected in this area during the fieldwork was taken into account in this study.

In addition to complaints made by residents to the City Hall telemarketing system, complaints made about this type of pest by residents who contact the Zoonosis Control Department (GCZ), whether by letter, oral statement, in person, by telephone or by e-mail, were also included. Also included in the study were different types of processes such as: petitions, requests from councillors, registrations at the City Hall call centres and demands from public bodies via official letters. The demands of the Zoonosis Control Management (GCZ) employees themselves were explored, usually made by the Health Agents, who replicate requests found during inspections. All the information was recorded in an EXCEL spreadsheet and analysed.

All the field collections were made once a month, as was the sum of the complaints that were closed within a month.

The monthly collections and recording of complaints took place between August 2014 and April 2015. Table 1 below shows the dates on which the field activities and subsequent laboratory identification were carried out.

Table 1. Dates of the monthly collections carried out during the study period at the RM-9/Medical School (Valparaiso) and Jardim Bom Pastor Retention Reservoirs.

Month	Valparaíso Swimming Pool	Bom Pastor swimming pool
August 2014	06/08	06/08
September 2014	23/09	23/09
October 2014	28/10	30/10
November 2014	25/11	26/11
December 2014	17/12	18/12
January/2015	21/01	22/01
February/2015	10/02	12/02
March 2015	25/03	26/03
April/2015	16/04	17/04
May/2015	12/05	13/05

Note: On 29/10/2014, the activity of vacuuming the resting boxes collected from the Faculdade de Medicina-RM9 swimming pool, Valparaíso, was carried out.

4.4 TIME AND CLIMATE ANALYSIS

4.4.1 Climate data

The climate data was obtained from SEMASA and specifically the climate data for the two reservoirs coincide, as they are in the same region, including the measuring station inside the RM-9/Medical School Reservoir (Valparaíso) and is called the Remote Transmission Unit (UTR-78).

The data captured by this station is recorded by the Elipse software (manufacturer Vector Engenharia e Sistemas de Alimentação de Dados), which was acquired by SEMASA.

4.4.1.1 Precipitation and Relative Humidity

Daily rainfall and relative humidity are climatological data that were analysed because it is known that on rainy days there is a carry-over of eggs and immature forms of Culex quinquefasciatus since the female of this species oviposits directly in the water and so any increase in flow caused by rain makes it easy for these forms to be displaced from their original breeding sites (Morais, 2005).

One factor that certainly affected the research is that the south-east of Brazil, of which the municipality of Santo André is a part, went through a period of drought in 2013 and 2014, which also affected 2015, being considered the worst drought in the region in 84 years (Beer & Thomas, 2014).

Relative humidity plays a role because, as well as causing a reduction in breeding sites due to evaporation on very dry days, it acts directly on the immature forms, desiccating them in the case of very low levels, as well as on the adult forms if there is a reduction in shelters, as in the case of extensive weeding.

4.4.1.2 Ambient Temperature and Monthly Average

Mosquitoes, like all insects, are pecilothermic and, as previously described, highly adapted to the tropics. It is therefore essential to measure the ambient temperature. On warmer days and months, there is a greater proliferation of immature and adult forms, while on colder days and months the opposite occurs, and mosquito deaths even occur (Forattini, 1962; Forattini, 2002).

4.5 STATISTICAL ANALYSIS

Data such as dominance, similarity, diversity, equity, richness, abundance and frequency were analysed and compared using different strategies for the different collection techniques used. Statistical tests were selected to demonstrate the validation of the indicators

used from a practical point of view. The choice of indices or statistical formulae depended on the demonstrations required, which were defined throughout the study as the results were processed (Bussab, 2006).

When analysing other scientific studies such as Natal (1991), Bracco (1998), Morais (2005), Vitor Neto (2006), Barata et al.(2007), Morais and Natal (2007), Silvério (2008), Silvério and Urbinatti (2011), it was observed that Culex quinquefasciatus reaches values close to 100 per cent in environments with water rich in organic matter and when we checked at the beginning of the laboratory identification and counting work, we realised that this fact would be repeated and so statistical concepts that dealt with interspecific relationships with other species such as dominance, similarity, diversity, richness and abundance were abandoned (Bussab, 2006).

Concepts such as fairness are embedded in the customer service script, which deals with questions that the technicians have applied and trained the telemarketers to ask in order to differentiate between the target pest, in the case of the common mosquito and the dengue mosquito, for example.

The script includes questions about the time of the bites, if there is any night-time discomfort, if the complainant hears buzzing, if they have seen mosquitoes, etc. However, the training was given years ago, there is a high turnover of female attendants, and there has been a change in the municipal administration as well as the contract with an outsourced company. Despite all this, the concept could not be applied because there were technical problems with the customer service system and residents began to access City Hall directly via the internet to make their complaints, which made it impossible to apply the script, so because of all this, this concept of equity was also abandoned (Bussab, 2006).

The statistical concept for this research was limited to frequency, not least because the scope of this work is to evaluate mosquito control. The identification of mosquitoes, although no less important, is more related to the scope of an academic Master's degree, while a Professional Master's degree has a more practical nature, to be applied in the service routine.

To choose a value to apply to the indices, the median was chosen.

4.6 INDEX CONSTRUCTION PROPOSAL

The indices proposed here are intended to be simple and easy to apply by pest, vector or zoonosis control agencies, as well as being based on easy-to-acquire instruments.

Indices, unlike indicators, which are much more complex, tend to be primary data and can form part of health or infestation indicators, or even ecological or social indicators, depending on the context in which they are used as an assessment tool (Belen, 2006).

The three infestation indices proposed here are quantitative, i.e. they reflect a count of insects on a variable while the complaints index is a qualitative index, as it is conditioned by a value judgement, which is the annoyance produced by the bites for the population and not the number of bites counted.

Meadows (1998) in Belen (2006) makes an interesting analogy between index and indicator. According to the researcher, the index would be the constant number in the mercury column of the thermometer and the indicator would be the fever, which is a symptom, a set of clinical signs. She says that the amount (and therefore usefulness) that the tool is used causes feedback and valorisation to occur. In the case of the thermometer, variables that can occur such as the ink on the number being erased, the calibration of the mercury, the influence of the ambient temperature, the malleability and expansion of the mercury are becoming less and less frequent, but there are still variables that can persist such as the vision of the person reading the thermometer. But much more sensitive to the variables cited as pitfalls will be the interpretation of the indicator. In this case, it is necessary to look at variables such as species, ambient temperature, metabolism and other clinical signs such as prostration in order to determine whether it is really a fever.

Indicators are in fact a model of reality, but they cannot be considered reality itself. However, they must be analytically legitimate and constructed within a coherent measurement methodology, being signals referring to complex events and systems (Belen 2006).

Based on the techniques used in the fieldwork, the following indices were constructed, described below:

4.6.1 Density of Immatures per Litre (DIL)

The Density of Immatures per Litre (DIL) index was proposed because there are differences between the sizes of entomological shells, ranging from 150 ml to 900 ml. Because of this, we looked for a standard size of a very frequent graduation that would fit in a container with the characteristics of being easy to transport and that would hold a large number of immature forms without them damaging each other due to lack of space in the container.

A quantity of one litre was arrived at and as soon as the water collection reached this level of graduation, the immature forms were counted, regardless of the number of shells (we try to follow the M.S. 2011 standard as much as possible) giving the direct result of the index. The index ratio is therefore the number of larvae per litre of water collected.

4.6.2 Index of Mosquitoes Aspirated per Minute (IMAM)

The Index of Mosquitoes Aspirated per Minute (IMAM) is calculated by counting the mosquitoes aspirated for five minutes by the Nasci aspirator and dividing the number found by five. It should be remembered that the IMAM is obtained individually, i.e. for each aspiration. When calculating a general IMAM for a site, such as a swimming pool where five aspirations have been made, the sum should be added up and divided by twenty-five (25) in this case. The IMAM can be reclassified, i.e. it is possible to make an IMAM for male mosquitoes and an IMAM for females. The ratio would be the number of mosquitoes collected by the number of minutes spent on each aspiration or it could be the sum of aspirations.

4.6.3 Index of Mosquitoes Aspirated per Resting Box (IMACR)

The Index of Mosquitoes Aspirated per Resting Box (IMACR) is calculated by counting the mosquitoes aspirated by the Natal and Marucci aspirator from inside each resting box individually and directly, i.e. each box will give one IMACR. The IMACR, like the IMAM, can be subdivided into male and female IMACRs. The ratio will be the number of mosquitoes aspirated per resting box.

The IMACR, unlike the other infestation indices proposed here (DIL and IMAM) seems to be just a count, but this is due to the fact that it undergoes many more variables than other indices that are conditioned to simple variables such as time and space. Firstly, the index cannot be considered simply a count because it can be broken down, as it was in this study, into the IMACR of males and the IMACR of females.

The IMACR calculated here should be the number of mosquitoes aspirated from the inside of Styrofoam boxes, 36 centimetres high, painted with black latex paint on the inside, with an exposure time of 18 hours, in an urban environment, with a wet cotton cloth attractant at room temperature.

Due to the high number of variables involved in this activity, it is more difficult to establish an index. In the discussion, we'll see that there are several variables that interact in the rest room activity.

As variables, we can mention the shape (in Burkot et al.'s (2013) work, what acts as the resting box is a screen stretched between two bamboo poles), the material of the resting box (Komar et al. (1995) used a fibreglass container while Van Bijllaardt et al. (2009) used clay pots); the material lining the inside of the box (Manda et al. (2011) used various materials, some impregnated with insecticides, while Pombi et al. (2014) used adhesive substances); the exposure time (Govella et al. (2011) compared it to the Human Landing Capture time); the host species (Govella et al. 2011); the mosquito attractant (Kweka et al. (2009) used various attractants such as boxes with odour baits that simulate human sweat and even cattle urine); whether the box is in an urban or rural environment (Barata et al. 2007, Sandhu 2013); if it is in desert areas (Van Bijllaardt et al. 2009) or areas with a high density of vegetation cover (Harbison et al. 2006, Burkot et al. 2013); with a number of competing breeding sites (Burkot et al. 2013); if it is in places with a high ambient temperature (Burkot et al.(2013)) or if there is a need to heat the resting box (Manda et al. 2011); if the box is in the home or peridomicile (De Meillon et al. 1967, Edman et al. 1997, Barata et al. 2007); the physiological state of the female mosquito (Braks & Cardé 2007, Sandhu et al. 2013); and the biological need, whether feeding, breeding or resting (Burkot et al. 2013).

Because of these variables, the IMACR is more difficult to fit into the space-time dimensions, unlike the DIL and IMAM.

4.6.4 Corresponding Complaint Index (CRI)

The Corresponding Complaint Index (CRI) consists of obtaining a database of requests from residents (in the case of Santo André, the software is called SIA - Sistema de Informação de Atendimento), the next step is to identify the target pest, georeference it, i.e. locate where the nuisance is coming from, In the case of mosquitoes, an application (Google Maps) was used to relate the address of the nuisance to the geographical position of the source of infestation, in this case the swimming pools, taking into account the flight radius of the mosquitoes, which was considered to be up to two kilometres (2 km). Thus, the monthly sum of these complaints will make up the IRC. The ratio will be the number of complaints within 2 kilometres of the study site per month.

It is advisable not to apply this index immediately, but rather to wait for the result of the inspection generated by the complaint because it could be discarded (for example, described as a mosquito and belonging to another species) or even increase the index (for example, the cognitive bias generated by the media about dengue, when the fear of the population "transforms" everything into a dengue mosquito, when in reality it is Culex).

Unlike infestation indices, which are quantitative, the IRC is a qualitative index, which is subject to greater subjectivity.

CHAPTER 5

RESULTS AND DISCUSSION

RESULTS

The results of the different collections carried out during the fieldwork are presented below.

5.1 MONTHLY FREQUENCIES OF IMMATURE MOSQUITOES COLLECTED IN THE VALPARAÍSO AND BOM PASTOR POOLS

Table 1 shows the number of immature forms (larvae and pupae) collected from the two pools during the fieldwork period from August 2014 to May 2015.

Table 1. Number of immature forms (pupae and larvae) of mosquitoes collected monthly in the Valparaíso and Bom Pastor Reservoirs from August 2014 to May 2015.

Month/ year	Valparaíso Swimming Pool						Bom Pastor Swimming Pool						Grand Total
	LV 1	LV 2	LV 3	LV4	LV5	Total	LBP 1	LBP 2	LBP 3	LBP 4	LBP 5	**Total**	
Aug/14	0	0	0	0	0	0	0	0	0	105	208	**313**	313
Sep/14	0	0	0	0	0	0	0	0	0	67	27	94	94
Oct/14	0	0	0	0	0	0	0	0	0	0	208	208	208
Nov/14	0	0	0	0	0	0	0	0	2	0	0	0	2
Dec/14	3	0	0	0	0	3	0	0	0	0	0	0	3
Jan/15	0	0	0	0	0	0	0	0	0	0	0	0	0
Feb/15	0	0	0	0	0	0	1	0	0	0	0	**1**	1
Mar/15	0	0	0	0	0	0	0	0	0	0	0	0	0
Apr/15	0	0	0	0	0	0	0	0	0	0	0	0	0
May/15	0	0	0	0	0	0	0	0	0	0	0	0	0
Total	3	0	0	0	0	3	1	0	2	172	443	618	621

As we can see, there were practically no immatures in the RM-9 Faculty of Medicine (Valparaiso) pool, with only 3 specimens collected near the spillway (see Photo 1 in Appendix A), where the water proved to be more polluted in that reservoir.

At the Bom Pastor Retention Reservoir, on the other hand, there were large numbers of larvae at points LBP 4 (see photo 26 in Appendix A) in the first two months and LBP 5 (see photo 27 in Appendix A) in the first three months after fieldwork began, which no longer occurred, probably due to anthropogenic intervention, such as mechanical cleaning and weeding, but mainly the application of larvicide.

Table 2 shows the differentiation between the stages of the immature forms in the larvae and pupae collected.

Table 2. Monthly number of immatures by developmental stage (pupae and larvae) collected in the Valparaíso and Bom Pastor pools from August 2014 to May 2015.

Month/year	Valparaíso Swimming Pool			Bom Pastor Swimming Pool			
	L3/ L4	Pupae	Total	L3/ L4	Pupae	**Total**	**Grand Total**
Aug/14	0	0	0	284	29	**313**	313
Sep/14	0	0	0	90	4	94	94
Oct/14	0	0	0	208	0	208	208
Nov/14	0	0	0	1	1	2	2
Dec/14	1	2	3	0	0	0	3
Jan/15	0	0	0	0	0	0	0
Feb/15	0	0	0	1	0	**1**	1
Mar/15	0	0	0	0	0	0	0
Apr/15	0	0	0	0	0	0	0
May/15	0	0	0	0	0	0	0
Total	1	2	3	584	34	618	621

In the Bom Pastor reservoir, the number of larvae was much higher than the number of pupae found (approximately 17 times more), as was the number of immatures compared to the other reservoir.

It can be seen that the number of pupae was greater than the number of larvae in the Valparaíso pool, but the number of samples was very small.

Graph 1 below illustrates this situation:

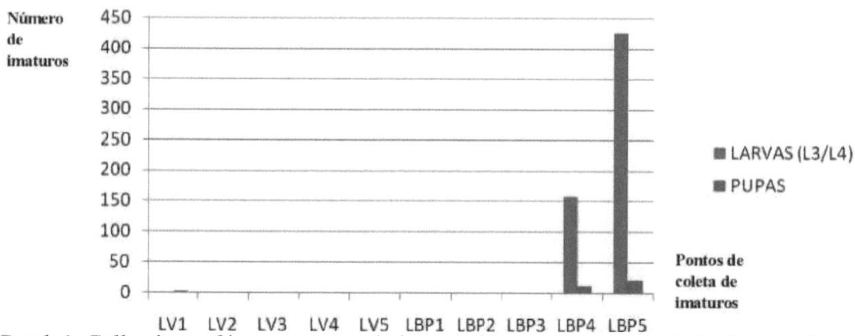

Graph 1. Collections of immatures carried out at the sampling points in the Valparaíso and Bom Pastor pools, Santo André municipality, between August 2014 and May 2015.

Graph 1 shows that only one sampling point in the Valparaíso Reservoir tested positive for immature forms, and only in small numbers, while in the Bom Pastor Reservoir four sampling points tested positive for immature forms, but only two in larger quantities.

Graph 2 below shows the number of larvae and pupae found in the RM-9 Faculty of Medicine Reservoir and the Bom Pastor Reservoir between August 2014 and May 2015.

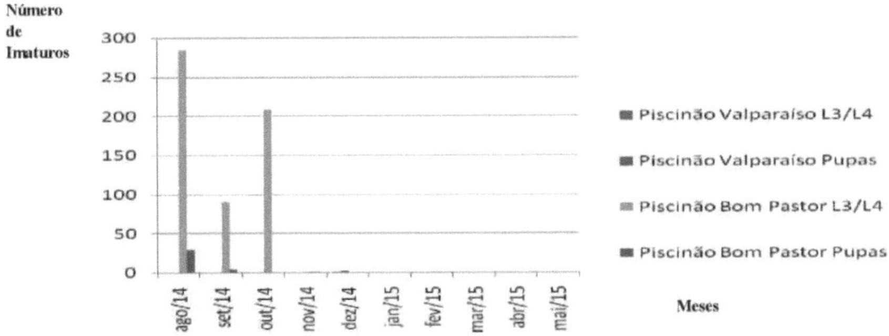

Graph 2. Monthly collections of differentiated immature forms carried out in the Valparaíso and Bom Pastor pools, Santo André municipality, between August 2014 and May 2015.

Graph 2 shows the numerical representativeness of the larval forms found in the Bom Pastor pond in relation to the pupal forms, as well as in the Valparaíso pond, the latter being much smaller in quantity.

Graph 3 below shows the number of pupae as a percentage and larvae found in both reservoirs.

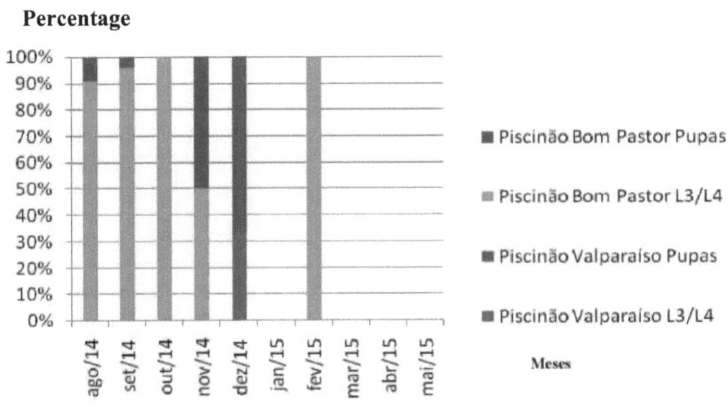

Graph 3. Percentages of immature forms differentiated into larvae and pupae in the Valparaíso and Bom Pastor swimming pools in the municipality of Santo André between August 2014 and May 2015.

Graph 3 shows that larvae were predominantly sampled between August and November 2014 and February 2015 in the Bom Pastor pool and pupae predominated in December 2014 in the Valparaíso pool.

5.2 MONTHLY FREQUENCIES OF ADULT MOSQUITOES ASPIRATED IN THE VALPARAÍSO AND BOM PASTOR SWIMMING POOLS

Table 3 deals with vacuuming activities carried out on the vegetation in the reservoirs.

Table 3. Number of adult culicidae separated by sex collected in the shelters of the sampling points of the Valparaíso and Bom Pastor pools from August 2014 to May 2015.

Month/year	Valparaíso Swimming Pool										
	Point 1		Point 2		Point 3		Point 4		Point 5		Total
	M	F	M	F	M	F	M	F	M	F	
Aug/14	0	0	1	0	4	0	4	2	4	0	15
Sep/14	4	1	0	0	0	0	0	0	0	0	5
Oct/14	5	0	2	1	0	0	0	1	0	0	9
Nov/14	1	0	3	0	11	2	32	19	3	1	72
Dec/14	3	1	4	0	0	0	15	4	0	0	27
Jan/15	0	0	1	3	0	0	7	4	0	1	16
Feb/15	4	2	4	0	0	0	11	4	3	0	28
Mar/15	1	0	0	0	5	1	0	1	9	0	17
Apr/15	1	0	2	0	1	0	12	3	0	0	19
May/15	1	0	8	2	0	0	4	7	0	0	22
Total	20	4	25	6	21	3	85	45	19	2	230

Month/year	Bom Pastor Swimming Pool											Total General
	Point 1		Point 2		Point 3		Point 4		Point 5		Total	
	M	F	M	F	M	F	M	F	M	F		
Aug/14	15	6	9	5	25	15	7	6	1	2	91	106
Sep/14	23	0	6	2	0	0	3	0	3	0	37	42
Oct/14	67	9	5	6	4	4	5	9	2	2	113	122
Nov/14	13	1	15	5	245	104	32	2	10	6	433	505
Dec/14	8	0	79	23	425	195	24	1	23	13	791	818
Jan/15	3	1	24	12	6	1	2	1	0	0	50	66
Feb/15	2	0	9	4	8	19	1	0	3	1	47	75
Mar/15	6	0	3	6	2	1	14	2	1	1	36	53
Apr/15	8	3	12	5	6	6	7	0	1	1	49	68
May/15	1	1	3	9	2	2	1	0	0	0	19	41
Total	146	21	165	77	723	347	96	21	44	26	1666	1896

As we can see, 1,896 adult mosquitoes were aspirated, with a greater predominance in the Bom Pastor Reservoir, with 87.86 % of the total sampling, with point ABP 3 (photos 30 and 31 in Appendix A) having hundreds of specimens captured in November and December that contributed to this result.

Graph 4 below shows the collections of aspirated mosquitoes in the Valparaíso swimming pool, separated by sex.

Number of mosquitoes

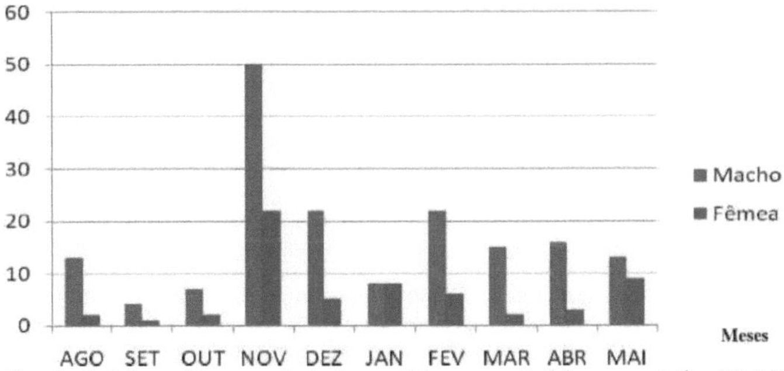

Graph 4. Mosquitoes aspirated monthly separated by sex at the RM 9 Medical School Reservoir, Valparaíso, Santo André, from August 2014 to May 2015.

We found that in the Valparaíso Reservoir the most prolific months were November, December and February, and in all months males predominated over females, not least because it is an external and circumscribed area. It is believed that this is because, as it is an urban area, the females enter the houses to feed and stay in the pools before fertilising, abandoning the site after mating.

Graph 5 below shows the mosquitoes separated by sex that were aspirated in the Bom Pastor swimming pool during the same period.

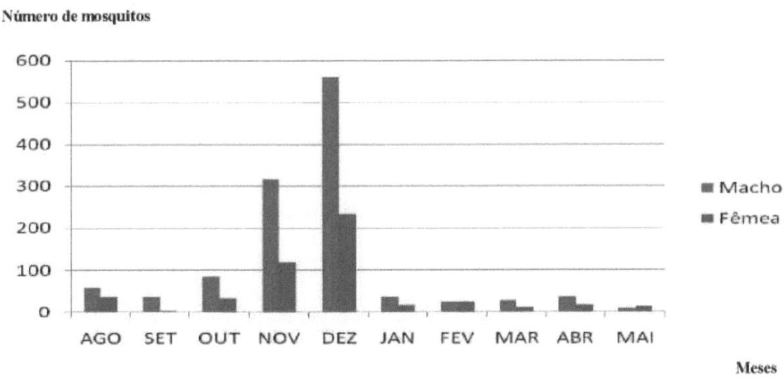

Graph 5. Mosquitoes aspirated monthly separated by sex in the Bom Pastor Retention Reservoir, Santo André, from August 2014 to May 2015.

What we can see that differs from the graph for the other reservoir is that the Bom Pastor pool, because it has collection points close to homes (see points ABP 3 and ABP 4, in photos 30 and 32, respectively, in Appendix A) in some months (February and May) there was a predominance of females at the suction points, but because it is also an outdoor area, in general there was a predominance of males. It should be noted that the number of mosquitoes on the vertical axis of the graph is in the hundreds, while in the other reservoir it is in the tens.

Graph 6 below gives an overview of aspiration in shelters in the two reservoirs throughout the study period.

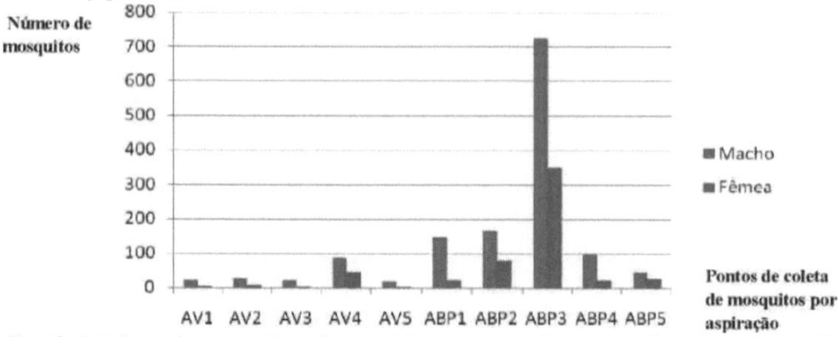

Graph 6. Mosquitoes aspirated monthly separated by sex at the aspiration points of the Valparaíso and Bom Pastor swimming pools, Santo André from August 2014 to May 2015.

Graph 6 shows that males predominated over females during all the months of fieldwork.

5.2 MONTHLY FREQUENCIES OF MOSQUITOES COLLECTED IN THE RESTING BOXES INSTALLED IN THE VALPARAÍSO AND BOM PASTOR POOLS

The results of the activity of aspirating mosquitoes from inside the resting boxes installed in the two reservoirs are shown in **Table 4** below.

Table 4. Number of culicidae separated by sex, aspirated monthly from the resting boxes installed in the Valparaíso pool and the Bom Pastor pool from August 2014 to May 2015.

Monthly Number of Adult Culicidae Collected from Resting Places																							
	Valparaíso Swimming Pool											Bom Pastor Swimming Pool											
Month	CR 1		CR 2		CR 3		CR 4		CR 5		**Total**	CR 6		CR 7		CR 8		CR 9		CR 10		**Total**	**Total General**
	M	F	M	F	M	F	M	F	M	F		M	F	M	F	M	F	M	F	M	F		
Aug	0	0	0	0	0	0	0	0	0	0	0	0	0	0	0	0	0	0	0	0	0	0	0
Set	0	0	3	0	0	0	0	0	1	0	4	0	0	0	0	0	0	0	0	0	0	0	4
Out	0	0	1	0	3	0	1	0	0	0	5	4	1	0	0	0	0	0	0	0	0	5	10
Nov	0	0	0	0	1	0	0	0	0	0	1	0	0	1	0	1	0	0	0	8	0	10	11
Ten	0	0	0	0	0	0	0	0	0	0	0	5	0	0	0	2	1	3	1	9	1	22	22
Jan	1	1	7	1	0	0	0	0	1	1	12	0	0	0	0	0	0	0	0	0	0	0	12
Feb	0	0	1	0	0	0	0	0	0	0	1	0	0	0	0	0	0	0	0	3	1	4	5
Sea	0	0	0	0	0	0	0	0	0	0	0	0	0	0	0	0	0	0	0	0	0	0	0
Apr	0	0	2	0	0	0	0	0	0	0	2	0	0	0	0	0	0	0	1	0	0	1	3
May	0	0	0	0	0	0	0	0	0	0	0	0	0	0	0	0	0	0	0	0	0	0	0
Total	1	1	14	1	4	0	1	0	2	1	25	9	1	1	0	3	1	3	2	20	2	42	67

We can see in the table above that the number of specimens collected (67 in total) was lower than the field aspiration technique and the number of mosquitoes prevailed in the Bom Pastor Reservoir over the Valparaíso Reservoir (42 and 25).

Graph 7 below shows the results of the resting boxes installed at the RM-9 Faculty of Medicine Reservoir (Valparaiso), separated by sex of the mosquitoes aspirated from the resting boxes.

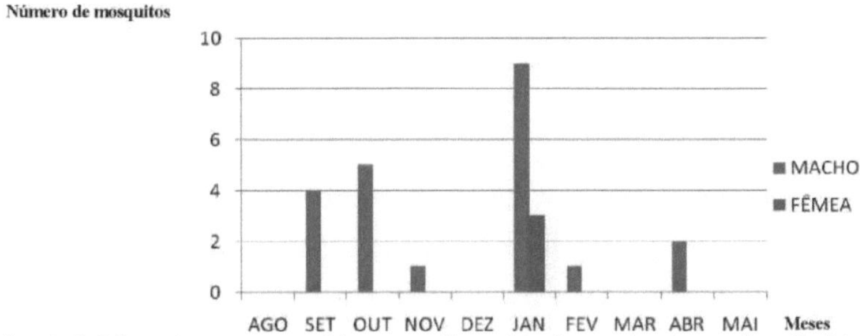

Graph 7. Mosquitoes separated by sex collected in the resting boxes installed in the RM 9 Medical School Reservoir, Valparaíso, Santo André from August 2014 to May 2015.

We observed a predominance of males collected and, with the exception of January 2015, an absence of females at the site.

The resting boxes that tested positive for females in January were close to the watchmen's quarters (see photo 16 in Appendix A), which corroborates their report of night-time disturbance.

Graph 8 below gives an overview of the results found from the aspirations of the resting boxes installed in the Bom Pastor Reservoir.

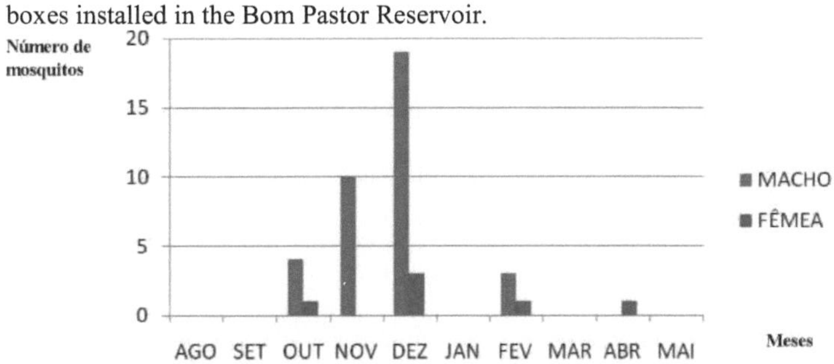

Graph 8. Mosquitoes separated by sex collected in the resting boxes installed in the Bom Pastor Retention Reservoir, Santo André, from August 2014 to May 2015.

What is striking about this graph is that, like the aspirations made in the vegetation, the presence of females is also noticeable here, probably due to the characteristics of the Bom Pastor Reservoir, which permeates homes.

In April 2015, the number of females collected using this technique predominated over the number of males, not least because the box installed is close to homes (see photo 38 in Appendix A). **Graph 9** below gives an overview of the results of the suction technique for the rest boxes installed in the two reservoirs.

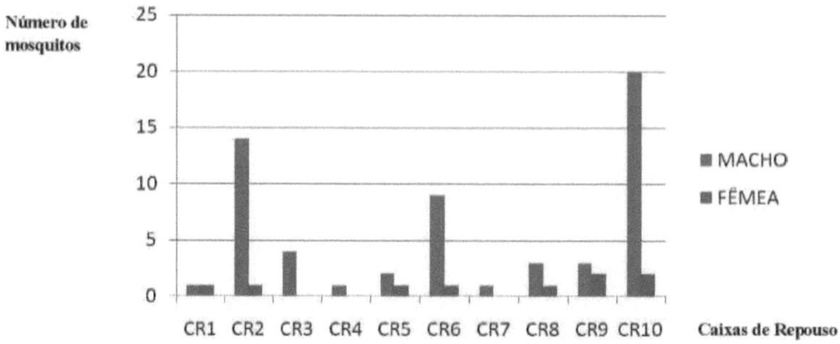

Graph 9. Mosquitoes separated by sex collected by resting box installed in the Valparaíso and Bom Pastor swimming pools, Santo André, from August 2014 to May 2015.

Overall, we found a prevalence of males over females using this resting box technique.

These activities showed the following results at the end of the study: the collection of pupae and larvae collected 621 specimens, the aspiration in the shelters collected 1896 specimens and the aspiration in the resting boxes collected 67 specimens. The reservoir located in Bom Pastor had more samples than the one in Valparaíso.

5.4 GRAPHICAL REPRESENTATION OF THE PROPOSED INDICES

The following graphs show the application of the indices proposed in this study.

Graph 10 shows the infestation rates calculated at the RM-9 Faculty of Medicine Reservoir (Valparaiso).

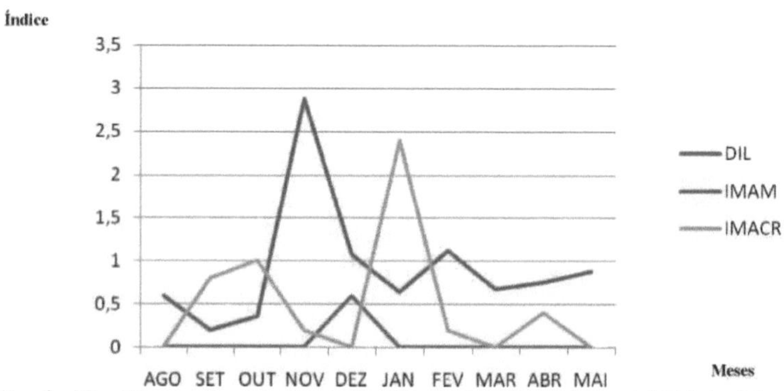

Graph 10 - Proposed mosquito infestation rates from August 2014 to May 2015 in the RM 9 swimming pool - Faculty of Medicine, Valparaíso, Santo André, where DIL is Density of Immatures per Litre, IMAM is Index of Mosquitoes Aspirated per Minute, IMACR is Index of Mosquitoes Aspirated from Resting Boxes.

We found that IMAM was higher in all months than DIL and in almost all months than IMACR, but IMACR was higher in September, October 2014 and January 2015, precisely the months in which mechanical cleaning took place in that reservoir.

Graph 11 below compares the same infestation rates with the Corresponding Complaint Index (CRI) for the region corresponding to Reservoir RM 9 - Faculty of Medicine (Valparaiso).

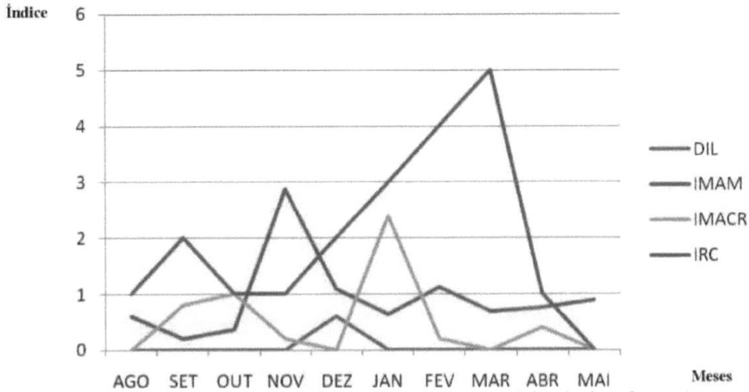

Graph 11 - Infestation rates and corresponding complaint index from August 2014 to May 2015 at the RM 9 swimming pool - Faculty of Medicine, Valparaíso, Santo André, where DIL is Density of Immatures per Litre, IMAM is Index of Mosquitoes Aspirated per Minute, IMACR is Index of Mosquitoes Aspirated from Resting Boxes and IRC is Corresponding Complaint Index.

What we noticed is that the IMACR, despite having lower numerical values than the IMAM, generally follows the IRC more closely, but in the month when it is highest (March 2015) the IMAM is the highest index.

Graph 12 shows the infestation rates in the Bom Pastor Reservoir:

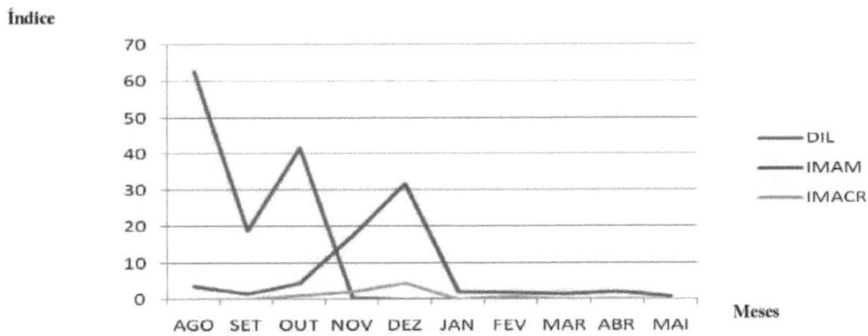

Graph 12: Proposed mosquito infestation indices from August 2014 to May 2015 in the Bom Pastor Retention Reservoir, Santo André where DIL is Density of Immatures per Litre, IMAM is Index of Mosquitoes Aspirated per Minute, IMACR is Index of Mosquitoes Aspirated from Resting Boxes.

Graph 12 shows that DIL is prevalent in the first three months due basically to two collection points in the Bom Pastor Reservoir, but due to probable anthropogenic intervention such as the application of larvicide, it is surpassed by the

IMAM, which shows higher numerical levels afterwards until the end of the proposed study period.

Graph 13 relates the IRC to the infestation rates in the Bom Pastor Reservoir.

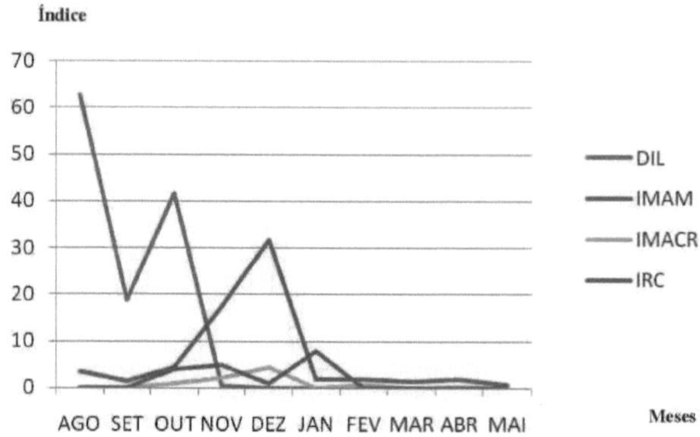

Graph 13 - Infestation rates and Corresponding Complaint Index from August 2014 to May 2015 in the Bom Pastor Retention Reservoir, Santo André where DIL is Density of Immatures per Litre, IMAM is Index of Mosquitoes Aspirated per Minute, IMACR is Index of Mosquitoes Aspirated from Resting Boxes and IRC is Corresponding Complaint Index.

Analysing **Graph 13**, the data appears to be disconnected, but if we take a closer look we can see that the highest IRC is in January 2015, while the highest IMAM and IMACR are in December, meaning that there is a delayed reaction from the population, which is normal in terms of the perception of discomfort. As the IMACR and IMAM fall, the IRC follows suit.

Graph 14 below shows the general infestation index, which means that the index was calculated for the two reservoirs together, which could be extrapolated to other breeding sites such as for the municipality and even larger regions as appropriate.

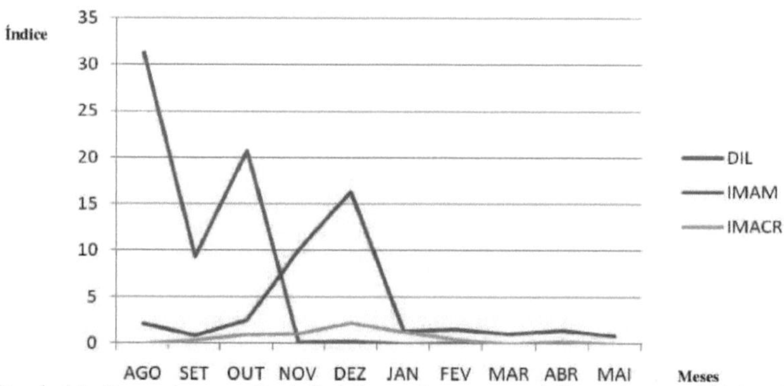

Graph 14. General mosquito infestation rates proposed from August 2014 to May 2015 in the Valparaíso and Bom Pastor swimming pools, Santo André where DIL is Density of Immatures per Litre, IMAM is Index of Mosquitoes Aspirated per Minute, IMACR is Index of Mosquitoes Aspirated from Resting Boxes.

We can thus see that hotspots that fall too far from numerical normality are reduced in importance even though they are still important, as was the case with DIL in this study. From December 2014 onwards, IMAM became more numerically significant.

Graph 15 shows the relationship between the general infestation indices of the two reservoirs and the general IRC (average of the two reservoirs' IRCs).

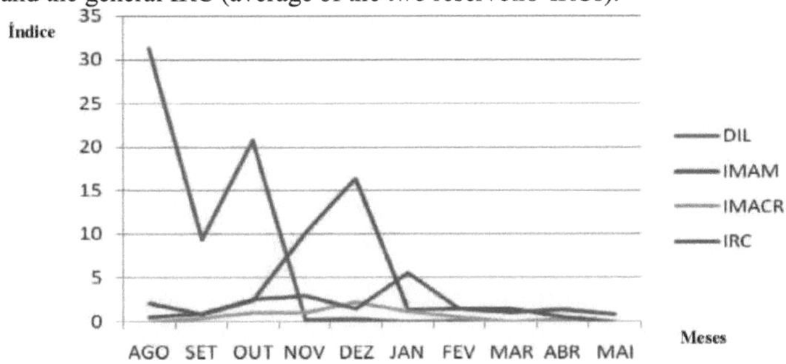

Graph 15 - General infestation rates and corresponding complaints from August 2014 to May 2015 at the Valparaíso and Bom Pastor Reservoirs, Santo André where DIL is Density of Immatures per Litre, IMAM is Index of Mosquitoes Aspirated per Minute, IMACR is Index of Mosquitoes Aspirated from Resting Boxes and IRC is Corresponding Complaint Index.

Graph 15 shows that the IRC does not have a direct relationship with the DIL, after all it is the adult forms that cause the nuisance, but the IRC does show an increase in relation to the higher numerical values indicated by the IMAM and IMACR indices, only in the following month.

The IRC (Corresponding Complaint Index) was based on 38 complaints from residents between August 2014 and May 2015, 20 of which corresponded to the Valparaíso Reservoir and 18 to the Bom Pastor Reservoir. The month with the highest number of complaints corresponding to the Valparaíso Reservoir was March, while the month with the highest number corresponding to the Bom Pastor Reservoir was January.

Graph 16 shows the mosquitoes collected by the aspiration technique and identified by sex in the RM-9 Faculty of Medicine Reservoir.

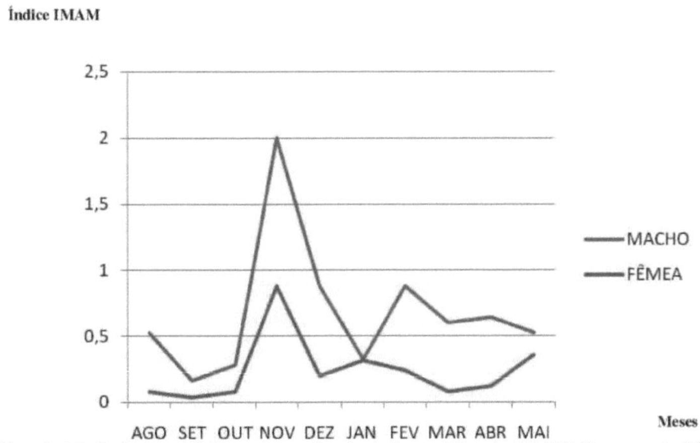

Graph 16. Indices of Mosquitoes Aspirated per Minute (IMAM) separated by sex at the RM-9 Faculty of Medicine Reservoir, Santo André from August 2014 to May 2015.

As we can see, there was a predominance of males, with the exception of January when the reservoir was mechanically cleaned, but only one more female was sucked up than the number of males.

Graph 17 shows the mosquitoes collected by the aspiration technique and identified by sex in the Bom Pastor Reservoir.

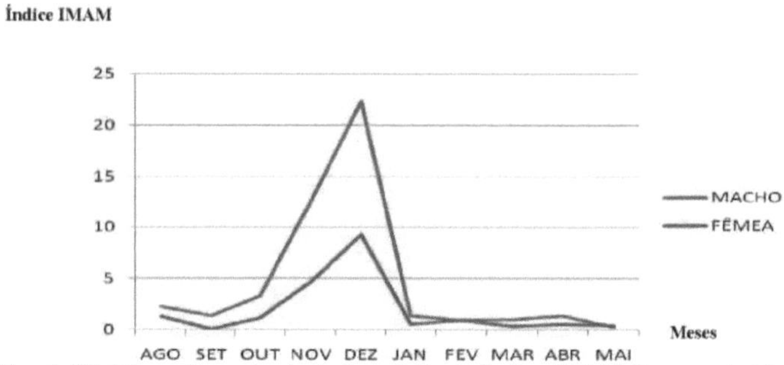

Graph 17: Mosquitoes Aspirated per Minute Index (IMAM) separated by sex at the Bom Pastor swimming pool, Santo André, from August 2014 to May 2015.

We found that in almost all the months the number of males prevailed over the number of females, with the exception of February and May, possibly due to their closer proximity to homes than the other reservoir.

Graph 18 shows the overall IMAM calculated on the results of the two reservoirs together.

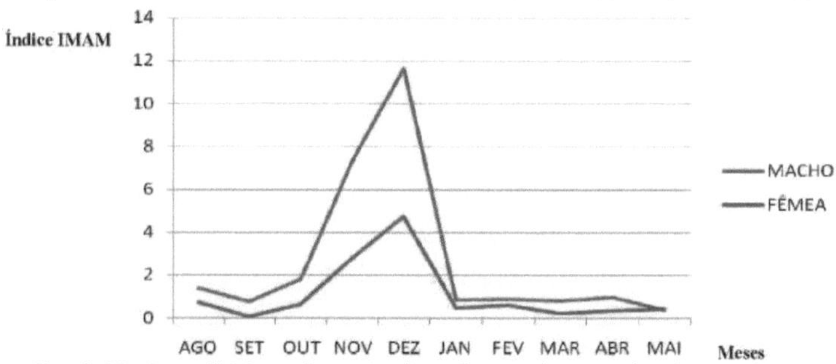

Graph 18. Overall Mosquitoes Aspirated per Minute Index (IMAM) separated by sex in the Valparaíso and Bom Pastor swimming pools, Santo André, from August 2014 to May 2015.

We found that the number of males seen generally exceeded the number of females in all the months and at all the suction points, probably due to the urban nature of the two reservoirs.

Graph 19 below shows the situation found in the resting boxes installed in the RM-9 Faculty of Medicine Reservoir.

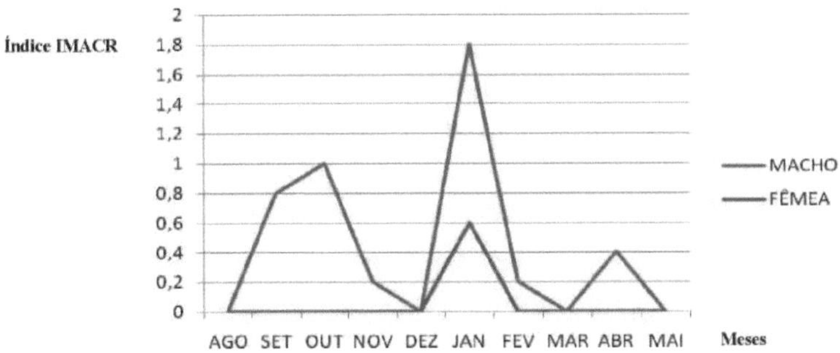

Graph 19. Indices of Mosquitoes Aspirated in Resting Boxes (IMACR) separated by sex in the Valparaíso swimming pool, Santo André from August 2014 to May 2015.

Graph 19 shows an almost exclusive predominance of male mosquitoes found in the resting boxes installed in the Valparaíso swimming pool, with January 2015 being the month in which the largest sample of both males and females was collected.

Graph 20 below shows the situation found in the resting boxes installed in the Bom Pastor Reservoir:

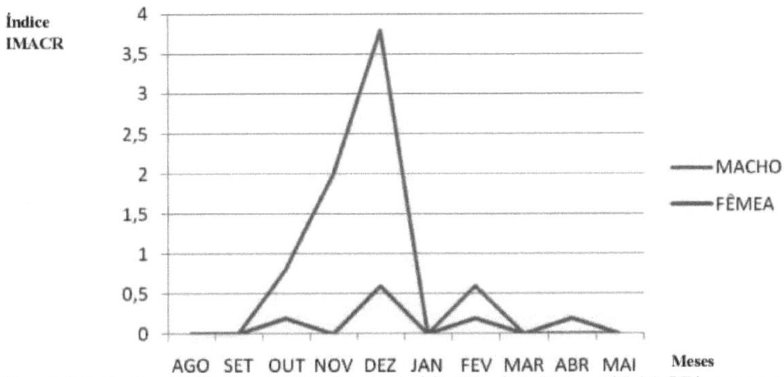

Graph 20. Indices of Mosquitoes Aspirated in Resting Boxes (IMACR) separated by sex in the Bom Pastor Swimming Pool, Santo André from August 2014 to May 2015.

As we can see, the male mosquitoes outnumbered the females in all the months in which the boxes tested positive, with the exception of April when the number of females exceeded the number of males.

Graph 21 shows the overall IMACR calculated for the two reservoirs.

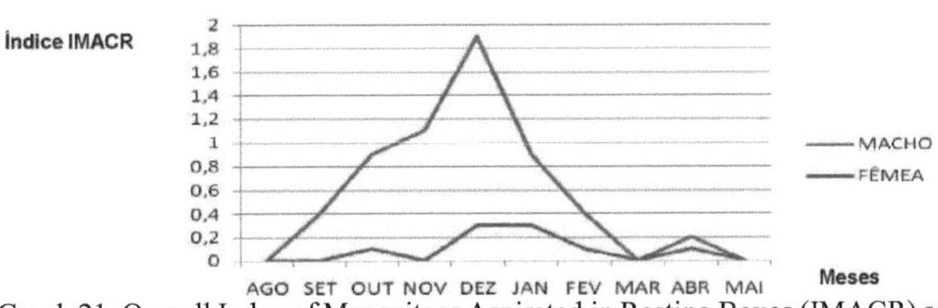

Graph 21. Overall Index of Mosquitoes Aspirated in Resting Boxes (IMACR) separated by sex in the

Valparaíso and Bom Pastor swimming pools, Santo André, from August 2014 to May 2015.

In general, the graph shows a predominance of males.

1.5 GRAPHIC REPRESENTATION OF THE COLLECTION ACTIVITIES CARRIED OUT

Graph 22 below is based on the count of individuals collected in the sampling work at the RM-9 Faculty of Medicine Reservoir (Valparaiso) and reflects the techniques used and the results presented.

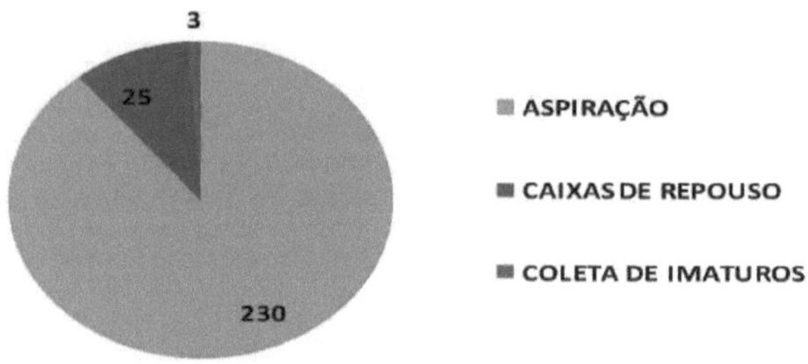

Graph 22. Number of individuals collected using different techniques in Reservoir RM 9 Faculdade de Medicina, Valparaíso, Santo André, from August 2014 to May 2015.

Aspiration using the Nasci aspirator was the technique that collected the most individuals, followed by the resting box technique and counting immatures.

Graph 23 below shows the count of individuals collected in the work sampling the Bom Pastor Retention Reservoir and reflects the results of the techniques used.

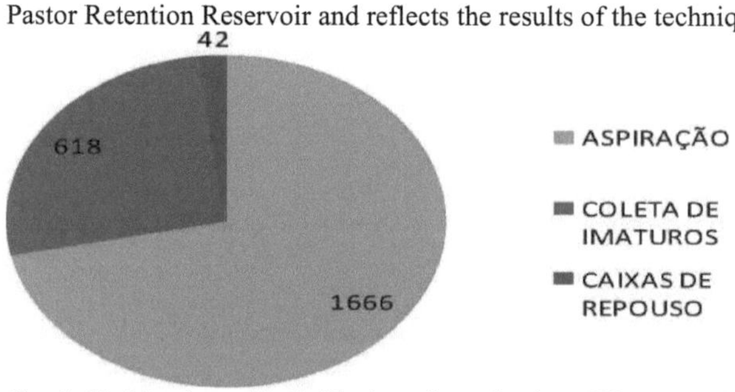

Graph 23. Number of individuals collected using different techniques in the Bom Pastor Retention Reservoir, Santo André, between August 2014 and May 2015.

Aspiration using the Nasci aspirator was the technique that collected the most individuals, followed by counting immatures and the resting box technique.

Graph 24 shows the final result of the different sampling techniques used in the two reservoirs.

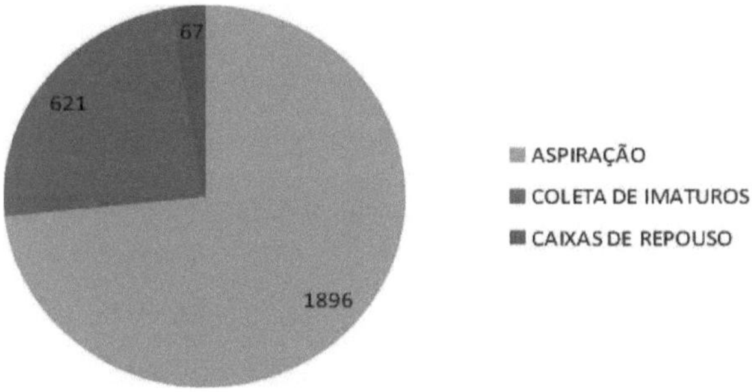

Graph 24. Number of individuals collected using different techniques in the Valparaíso and Bom Pastor Reservoirs, Santo André, between August 2014 and May 2015.

In general, aspiration using the Nasci aspirator was the technique that collected the most individuals, followed by counting immatures and the resting box technique.

1.6 AVERAGE TEMPERATURES AND RAINFALL RECORDED IN THE STUDY AREAS

As a complement to the study, climate data was obtained from the area and period of the experiment, as this information helps to interpret the entomological findings. Average monthly temperatures are shown below.

Table 2. Calculated average monthly temperatures for the location and period of fieldwork.

Aug/14	17,54 °C
Sep/14	19,24 °C
Oct/14	20,01 °C
Nov/14	20,53 °C
Dec/14	21,69 °C
Jan/15	24,11 °C
Feb/15	22,04 °C
Mar/15	21,54 °C
Apr/15	20,05 °C
May/15	17,99 °C

Source: CECOI/SEMASA/PMSA

According to the information gathered from RTU 78 (Remote Transmission Unit), which covers the region of the two reservoirs, August was the coldest month while January was the warmest. The average monthly rainfall is shown below.

Table 3. Average daily rainfall (in millimetres) per month calculated for the location and period of fieldwork.

Aug/14	1,28
Sep/14	2,22
Oct/14	0,27
Nov/14	2,98
Dec/14	4,80
Jan/15	7,66
Feb/15	9,46

Mar/15	6,05
Apr/15	1,79
May/15	2,65

Source: CECOI/SEMASA/PMSA

According to the information gathered from RTU 78 (Remote Transmission Unit), which covers the region of the two reservoirs, October was the month with the least rainfall, while February was the month with the most.

1.7 ANTHROPOGENIC AND NATURAL VARIABLES

This work deals with the development of infestation control indices and in order to discuss what is considered to be control, it is essential to look at the factors that interfere with mosquito proliferation, the so-called variables that can be many, but which can be grouped into natural ones that are the responsibility of nature and anthropic ones that are caused by human action. Among the natural variables, there are a large number that interact, as Urbinatti (2000) points out, and they can be biological (predation, parasitism, competition, etc.), physical-chemical (water pH, dissolved oxygen supply, presence of minerals, water conductivity, turbidity, etc.) or climatological.

For this study, we chose variables that certainly influence mosquito population growth, although we recognise that they are not the only ones, but their importance is undeniable: rainfall, relative humidity and ambient temperature.

We used the following situations as constraints:

In terms of ambient temperature, monthly average temperatures below 20° C were considered for the immature forms, based on Forattini (2002), who states that the optimum temperatures for larval development are above 24° C, so it was considered that although a water temperature in the 20° C range is not lethal for the larva, it is already a limiting factor for its development and that we are considering the monthly average, it means that there were not a few periods in this more uncomfortable range.

For the limiting temperature of the adult forms, we chose below 10°C based on experience with the dengue mosquito (Aedes aegypti), which despite being another species is a mosquito that is much more routinely collected by government agencies, which allows us to extrapolate to Culex and as we have already collected Aedes at temperatures of 8°C, we consider 10°C to be a reasonable measure for Culex as a limiting factor, considering its cosmopolitan character.

Rainfall is a carrying factor for larvae according to the work of Guanasekaran et al. (2002), Morais, (2005), Morais et al. (2007), inferring in larval sampling as a restrictive factor, although as it is a dynamic factor, the opposite can happen, such as the carrying of larvicides and other substances toxic to larvae, acting as a positive factor for proliferation (Victor Neto, 2006), but we consider it to be a restrictive factor. This factor was taken into account when precipitation occurred up to five days before the collection of immatures and all immature forms were carried away on the day of the precipitation and then counted from an oviposition at the site the following day and the minimum time of the cycle until larvae developed in stage 3 (L3), which is the minimum for counting in our proposed work.

In the case of adult mosquitoes, rainfall was considered to interfere only on the day of collection, since the pools studied are in an urban area that offers a large number of shelters that are safer from the rain than the open areas of the pools.

As for relative humidity, although when it is very low it is a factor that reduces breeding sites, it was not considered to interfere with the immature forms because they are in the aquatic phase of the mosquito life cycle.

According to Consoli and Oliveira (1994), air humidity is in the optimum range when it is above 80 per cent for Aedes mosquitoes and above 70 per cent for Culex mosquitoes, but adult mosquitoes easily protect themselves from this factor by hiding under vegetation where they find an ideal habitat. Values below 50% were considered an interfering factor in this study, but only in months when the reservoirs were weeded.

Rainfall interfered with the results of the samples throughout the study period, considering the conditions. **Appendix E** for daily rainfall shows that despite the dry period suffered by the south-eastern region of Brazil, rainfall occurred close to the days of the harvest.

The ambient temperature may have interfered in the colder months of August, September and May, contributing to a further decrease in the larval population, and the Relative Humidity may have interfered in the months of September and March. **Graph 4 shows** a decrease in the adult population compared to the previous month. Rainfall interfered in all months in the Bom Pastor reservoir, in the same way as the Valparaíso reservoir, after all both reservoirs are in the same region. So how can we explain the situation shown in **Graph 2** where we collected a large number of larvae in the hundreds per litre in the Bom Pastor pool?

The explanation is that in this case the rainfall acted more as a positive factor, bringing nutrients and oxygen to the larvae, than a negative one. The fact is that, particularly in the places where the samples were taken, this reservoir is not lined like the Valparaíso reservoir, and the shape of the reservoir helps to explain this. The fact that this reservoir has no circumscribed boundaries, forming meanders between the houses and streets, means that in the drier months the reservoir evaporates and due to the characteristics of the terrain it forms depressions where water accumulates and in the higher places it disappears, forming large puddles of water where there is no run-off, and the rainfall is insufficient for the water to flow. According to Forattini (2002) the water must be drained within 24 hours to prevent the formation of larvae. This happens in the Valparaíso pool, but not in the Bom Pastor pool.

The temperature may have had an influence in terms of containing proliferation a little; if it had been warmer, the number of larvae would have been even higher in August and September. In May, despite being a cold month, the influence of the ambient temperature was not taken into account because there were no larvae being collected in the previous months.

The Relative Humidity must have had a slightly greater influence, but as it is linked as a limiting factor to an anthropogenic activity, weeding, since with vegetation the adult forms are easily protected against low humidity, this will be discussed later. See photos 3 and 4 in Appendix B Weeding illustrating the situation in this pool.

Next we'll discuss anthropogenic activities, of which there are many, but we've chosen three as the most influential.

Mechanical cleaning with tractors, lorries and topcats carried out by third-party companies (see photos 1 and 2 in Appendix C Mechanical Cleaning) in itself, by removing the mud and sludge, causes a large die-off of larvae.

Weeding removes the natural harbours of adult mosquitoes, which has meant that the two actions alone have kept the mosquito population at relatively low levels.

In the months in which mechanical cleaning took place, there was an increase in the index of resting boxes (IMACR), which was close to the IMAM, which is the index of mosquitoes sucked up from the vegetation, most likely because of the agitation and dust collection, after all there was no undergrowth to hold the dust, and the mosquitoes sought shelter in the resting boxes.

The application of larvicide by ECRVAS/GCZ in April was not considered a major factor because practically no larvae were found in the previous months.

When analysing anthropogenic activities in the Bom Pastor Reservoir, we have to bear in mind that it has a different physical nature to Valparaíso, because it is like several environments in its extension and why not say different ecological niches.

When we receive a cleaning or weeding report, or even a disinsection report, it's important to know the exact location where the action was carried out, so that we don't make incorrect assessments.

To try to better understand the actions, we will try to identify the sampling point.

The mechanical cleaning carried out in August probably didn't reach points LBP 4 and LBP 5 (see **Table 1**) because they continued to be positive. Just to give you an idea of the diversity of environments in this pool, in Mechanical Cleaning Appendix C, photo 3 corresponds to Sampling Point LBP1, photo 4 to ABP4 (suction) and photo 5 to ABP 2 (also suction).

Weeding has certainly contributed to the reduction in adult forms and although it is done more constantly in this pool, it is more sparse and occasional. For example, in October it was only done in ABP 2 on the initiative of an employee of a private landscaping company. In November, depending on where it was carried out, it had no major short-term consequences for the proliferation of mosquitoes.

With regard to the application of larvicides and adulticides, when analysing Appendix D - Disinsection, the applications of larvicide based on Bacillus sphaericus toxin crystals on 12 November and then BTI on 08 April, it was not possible to assess whether it worked because these points were negative for larvae throughout the period.

On the other hand, the application of Bacillus sphaericus on 24 November seems to have been crucial for the disappearance of the immature forms by May 2015. Studies by Lopes et al. (2010) state that the average persistence period of these biological compounds is between 7 and 14 days, but Bacillus sphaericus has its efficiency reduced if the water is not moving and if there is sunlight. Pontes et al. (2005) state that these compounds, particularly BTI, work for up to 35 days if left in the shade, decreasing to 8 days if left in sunlight.

The use of insecticides when methyl pirimiphos was used to kill adult mosquitoes was ineffective, firstly because it was used as a spray with a 1200 PSI four-stroke motor pump, which is suitable for crawling insects, but not for winged insects, where a UBV or fogger should be used, and even then only for immediate effects. The ineffectiveness of this disinsection on 15 December can be seen in **photo 31** of 18 December in Appendix A. What's more, there was no rainfall from 15 to 18 December, as can be seen in Appendix E Precipitation. The use of foliar adhesives is not permitted in urban areas, but only in agriculture. However, the use of pirimiphos methyl may have potentiated the larvicidal action because when it runs off the vegetation it reaches the water bodies and the water in the Bom Pastor pool is relatively clean at some points and according to Sales and Hervy (1974) this product persists for up to 11 weeks in drinking water, in a study carried out in Burkina Faso, with a good residual.

1.8 ANALYSING SEASONAL ASPECTS FOR CONTROL INTERVENTIONS

After considering the variables, to carry out this analysis we applied the indices proposed in the tables below, where we also used the median.

Table 4. Infestation index values and complaints corresponding to the RM-9 Faculty of Medicine Reservoir (Valparaiso) from August 2014 to May 2015 and the median of these values.

	DIL	IMAM	IMACR	IRC
AUG	0	0,6	0	1
SET	0	0,2	0,8	2
OUT	0	0,36	1	1
NOV	0	2,88	0,2	1
TEN	0,6	1,08	0	2
JAN	0	0,64	2,4	3
FEB	0	1,12	0,2	4
MAR	0	0,68	0	5
APR	0	0,76	0,4	1
MAY	0	0,88	0	0
Median	0	0,72	0,2	1,5

Analysing the table above and based on the median and the LID, the Valparaíso pool did not prove necessary for larval treatment during the study period, showing that it was not an important breeding ground.

The IMAM showed that 0.72 mosquitoes aspirated per minute is already a control measure that should be carried out in November, December, February, April and May. More physical measures such as weeding and mechanical cleaning seem to be sufficient, and aspiration with a Nasci Vacuum Cleaner can be used to control these adult forms.

The IMACR showed a median of 0.2, i.e. as 5 resting boxes were installed in this reservoir, any mosquitoes found would merit special attention for control, i.e. in the months of September, October, November, January, February and April. The same recommendations follow for IMAM.

Analysing the IRC, we see that with this median any complaint about mosquitoes deserves to be addressed. The data for March may have a cognitive bias, because the focus of the breeding sites is probably not the Valparaíso swimming pool. Perhaps it would be more accurate if we reduced the radius of 2 km considered, because this month all the complaints were more than 1 km, not to mention that the media and the reality of the dengue epidemic this month make the population frightened about dengue and Culex rides on this fear.

Graph 11 illustrates these indices discussed above.

Table 5. Infestation index values and complaints corresponding to the Bom Pastor Retention Reservoir from August 2014 to May 20015 and the median of these values.

	DIL	IMAM	IMACR	IRC
AUG	62,6	3,64	0	0
SET	18,8	1,48	0	0
OUT	41,6	4,52	1	4
NOV	0,4	17,32	2	5
TEN	0	31,64	4,4	1
JAN	0	2	0	8
FEB	0,2	1,88	0,8	0
MAR	0	1,44	0	0
APR	0	1,96	0,2	0
MAY	0	0,76	0	0
Median	0,1	1,98	0,1	0

Table 5 shows that with a median LID of 0.1, every larva found requires a control measure. The months indicated for larvicide control are August, September, October, November and February.

The IMAM showed a median of 1.98 and the recommended months for adult control would be August, October, November, December and January. As larval control proved to be much more effective and practical when applying larvicide to water, controlling this aquatic stage of mosquitoes restricts the adult mosquito population.

As with the other reservoir, the IMACR median showed a low result, and the months indicated for mosquito control would be October, November, December, February and April. The same IMAM recommendations apply here, with the novelty of looking at the month of April.

Graph 13 illustrates the situations described above.

Table 6. Infestation index values and complaints corresponding to the RM-9 Faculdade de Medicina (Valparaíso) and Bom Pastor reservoirs from August 2014 to May 20015 and the median of these values.

	DIL	IMAM	IMACR	IRC
AUG	31,3	2,12	0	0,5
SET	9,4	0,84	0,4	1
OUT	20,8	2,44	1	2,5
NOV	0,2	10,1	1,1	3
TEN	0,3	16,36	2,2	1,5
JAN	0	1,32	1,2	5,5
FEB	0,1	1,5	0,5	1,5
MAR	0	1,06	0	1,5
APR	0	1,36	0,3	0,5
MAY	0	0,82	0	0
Median	0,15	1,43	0,45	1,5

Table 6 shows a result of the two reservoirs and this joint analysis is important because public managers have to deal with a shortage of resources, both human and material, and it is common

practice to group common establishments together and choose a period to pay attention to that type of establishment, such as a disinsection schedule for schools, health centres, etc. all at the same time.

That said, with a median value of 0.15, the months recommended by DIL for the two pools would be August, October, September, December and November, in order of importance for larval control. As the treatment of choice for control is larvicide, these are the right months to combat the common mosquito.

If the manager chooses to make a "fog" around the pools, which would be a more extreme measure, the most suitable months according to IMAM would be August, October, November, December and February. If the IMACR were to be followed, the values would be different, largely due to the anthropogenic activities carried out in the reservoirs, and the months would be October, November, December, January and February, i.e. the hottest months, which means that mosquitoes look for the resting place because of the heat.

1.9 PHYSIOLOGICAL STATE OF THE FEMALES IN THE SAMPLE

The samples of the adult forms showed a greater predominance of males than females in both the vacuuming technique in the shelters and the vacuuming of the resting boxes, but the females were identified in terms of their physiological state in order to get an idea of their biological behaviour.

Table 7 below shows the results found:

Table 7. Number of female mosquitoes according to their physiological state aspirated in the RM-9 Faculty of Medicine Reservoirs (Valparaiso) and Bom Pastor Retention Reservoir from August 2014 to May 2015.

	Valparaíso Swimming Pool			Bom Pastor Swimming Pool			Grand Total		
Month/ year	Empty	Pregnant women	Ingurgitated	Empty	Pregnant women	Ingurgitated	Empty	Pregnant women	Ingurgitated
Aug /14	1	1	0	28	6	0	29	7	0
Sep /14	0	1	0	1	1	0	1	2	0
ont/ 14	0	2	0	19	11	1	19	13	1
Nov /14	16	6	0	86	32	0	102	38	0
Dec /14	1	4	0	159	76	1	160	80	1
Jan/ 15	9	2	0	8	7	0	17	9	0
Feb /15	2	4	0	15	10	0	17	14	0
Mar /15	2	0	0	8	2	0	10	2	0
Apr /15	2	1	0	6	8	1	8	9	1
Ma y/15	6	3	0	9	3	0	15	6	0

Grand Total	39	24	0	339	156	3	378	180	3

Analysing the table, we can see that the number of empty females is higher than the number of pregnant females, showing that the reservoirs function as a place for adults to emerge, so many young females will show endophilia in the exercise of obtaining blood repasts when they are fertilised by males, despite the fact that males are also endophilic, but their habit of entering homes is due to the search for females and not due to feeding.

Another piece of information we get from the table is the low number of females.

All the results collected refer to the Culex quinquefasciatus species.

CHAPTER 6

DISCUSSION

Monitoring indices for immature and adult mosquitoes were constructed and validated.

During the course of the work, four monitoring indices were drawn up: the Immature Density per Litre (DIL), the Mosquito Aspiration Rate per Minute (IMAM), the Mosquito Aspiration Rate in Restrooms (IMACR) and the Corresponding Complaints Index (IRC).

According to Kluck et al. (2002), Philippi & Malheiros (2012) indices are one of the numerical expressions of an indicator that is a measure of activity.

To this end, the characteristics of a good indicator are usefulness (decision support), feasibility (availability of data), simplicity (easy to understand), reliability (represents what is happening), relevance, validity, sensitivity, discriminateness, comprehensiveness and ethics (Kluck et al. 2002).

In terms of usefulness, the indices are aimed at pest and vector control managers, they are feasible if they are publicised in official media outlets, they are extremely simple, and simple mathematical operations such as addition and division are all that is needed to calculate them.

In terms of reliability, i.e. the data must be reliable, capable of providing the same result if measured by different people in different environments and at different times, under similar conditions, the LID (Immature Density per Litre) proposed here is even more reliable than the index proposed by the Ministry of Health (2011), which is the DLP because it uses the ratio of the number of larvae per sample or "shell". This "conchada" comprises the capacity of the entomological shell, which is understood to be 350 ml, but on the market there are shells ranging from 150 to 900 ml, while the LID denominator is 1 litre, which is the most common measure of volume worldwide.

The indices are relevant in their very wording, they are valid in the sense that they do not reflect any parallel phenomenon, but they allow the interpretation of the data to be improved in an evaluation of the situation in search of a greater truth (Kluck et al. 2002).

Sensitivity, which according to Kluck (2002) is the degree to which it is able to identify all cases where there are problems in the current treatment of cases, the indices are sensitive, but there may be variation depending on the interpreter's level of knowledge and experience.

To increase the degree of sensitivity, it is recommended not to check the indices in isolation.

Also according to Kluck (2002), the proposed indices seem to fulfil their role in terms of discriminateness, which is the power to reflect different epidemiological and operational levels, even between areas with different particularities.

As for comprehensiveness, which is the synthesis of as many conditions or factors as possible that affect the situation of what you want to describe, it can be said that the indices proposed here are comprehensive because they touch on the situation of the nuisance population (mosquitoes) and the nuisance population (residents).

The methods for collecting adult mosquitoes by aspiration in shelters and resting boxes were validated, and the collection of immature mosquitoes for entomological monitoring purposes in swimming pool areas was validated by analysing the proposed indices.

In the case of flying insect control, preference and suitability is best given to larval control, when it is much easier for the active ingredient in the formulations to reach the target species. Dias (2000), when outlining a general overview of endemic disease control, is one of the authors who attests to this by casting doubt on the fogging technique, thus attributing more of a pyrotechnic and spectacle effect than effectiveness to this practice.

Freitas et al. (1996), in addition to citing numerous authors who choose to use both synthetic and natural larvicidal substances, carried out an experiment with a previously nematocidal substance such as ivermectin for the control of culicidae larvae and Alves et al. (2011) found that this substance had greater residual power than the traditional insecticides and larvicides tested in the experiment, such as pyrethroids and organophosphates.

Natural variables such as daily rainfall, average ambient temperature and relative humidity were analysed with SEMASA and anthropogenic variables such as mechanical cleaning, weeding, larvicide and adulticide application with DAEE, SEMASA and GCZ. It was found that LID is the index that is most altered by anthropogenic variables, a fact mentioned by Forattini (2002) when he mentioned the carrying of immatures by the torrent.

Oliveira et al. (2007) stated in a study in the city of Toledo (PR) that there was no correlation between Aedes aegypti infestation and relative humidity.

Garcia et al. (2010) carried out research in three Colombian cities with very different altitudes and average temperatures and observed that Culex quinquefasciatus adapts to higher and colder regions. Statistical analyses found no significant differences in mortality rates, survival rates and life expectancy in mosquitoes from the three cities chosen.

When Wright (1901) collected mosquito larvae, he realised that they were still alive under crusts of ice in Aberdeenshire (United Kingdom) at 32°F, which would correspond to 0°C ambient temperature. When he took them back to his laboratory on the morning of 14 February 1901, with the ambient temperature at 29°F, i.e. - 1.67°C, he was surprised to see that in a jar of water with an ice crust on the surface, there was a larva swimming freely at the bottom of the jar.

Tsuda et al. (2012) found that the Culex inatomii species increased after the tsunami in Japan in March 2011, mainly in pools with brackish water, comparing regions affected by the tsunami with unaffected regions. In this study carried out in Miyagi Prefecture, the authors found that the number of positive breeding sites with larvae increased from 2.7 per cent in June to 79.5 per cent in August.

In the immature collection activity, the overall number of larvae was higher than the number of pupae.

The most common method for collecting samples is by immersing shells in water collections suspected of being breeding grounds for mosquitoes, especially Culex quinquefasciatus, as attested to by Urbinatti (2000) in the Tietê Ecological Park and Morais (2005) in the Pinheiros River, even finding that this species is exclusive to this location. This collection method is known as "dipping" and there can be variations in the number of shells and the volume of water removed, with or without larvae.

Dibo et al. (2011) used the volume taken in 10 samples (scoops) of 625 ml each, totalling 6.25 litres per sampling point. De Paula & Gomes (2007) took around 30 scoops of 500 ml each, totalling 15 litres per ecotope or sampling point, but only considered positive scoops, i.e. those containing larvae or pupae inside.

Suguimoto & Natal (2009) found at a hydroelectric dam in Ourinhos (SP) that the culicidae fauna varied greatly in species richness and diversity, comparing data before and after the dam was filled, with a decrease in species in the post-flood period with the recolonisation of new species.

De Paula & Gomes (2007) found in a study of the Porto Primavera Dam in the municipality of Rosana (SP) that the number of species of the Culex genus increased in this dam after it was filled, which corroborates the fact that its construction was considered an environmental disaster and that it is considered the third most inefficient hydroelectric plant in the world.

Bueno-Mari et al. (2010) collected larvae for more than twenty years (19872007) in Valencia, Spain, mainly to draw a parallel to the profound environmental transformations that the region has undergone due to economic activities in the last century. This diversity of environments favours the proliferation of different species of culicidae depending, among other things, on the physical and chemical characteristics of the specific larval biotopes, as well as the presence of suitable hosts in the vicinity and the local climate. In this regard, various authors have proposed the use of culicidae as bioindicators of forest degradation and, in general, of anthropogenic pressure on various natural environments (Montes, 2005). However, studies on the use of mosquitoes as organisms sensitive to variations in environmental quality are scarce. In this study by Bueno-Mari et al. the methodology was "dipping" (350 ml clamshells) to collect larvae and in the case of little water, pipetting was used. However, the collection method was not based on the number of scoops or the amount of water removed, but rather on the time taken, spending 10 (ten) minutes per biotope (sampling site). In Spain, the largest number of larvae occurs between March and October. Adults were collected using a hoover and the shovel technique. *Culex quinquefasciatus* was the most frequent species of the 47,131 specimens collected and also the species with the greatest ecological plasticity.

In the adult sampling activities, mosquitoes were identified by sex and the number of males predominated over females at all sampling points, with the number of males totalling 70.89% in the shelter vacuuming activity and 86.57% in the resting box vacuuming activity. The physiological state of the females was also identified and the number of empty females predominated over the number of pregnant and engorged females. Morais & Marrelli (2006) also collected more males in aspirations carried out on the banks of the River Pinheiros in São Paulo.

Aspirations of adults in the shelters were most productive in November at the RM-9 Faculty of Medicine Reservoir (Valparaiso) and in December at the Bom Pastor Reservoir. The aspiration was carried out for five minutes and the IMAM in this study had a denominator of 5, but this can vary, as in the study by Dos Anjos & Navarro Silva (2008) in which they used the Nasci aspirator for thirty minutes at each aspiration site, in which case the denominator would have a value of 30.

Morais & Marrelli (2006) in aspirations carried out on the banks of the River Pinheiros in São Paulo spent five minutes at each aspiration point, as in this study, while De Paula & Gomes (2007) carried out aspirations for 15 minutes at the Porto Primavera Reservoir.

The IMACR, relating to the resting boxes, is more difficult to operationalise and showed very low numbers, representing only 3.53% of the sample universe compared to the IMAM and 2.59% of the overall sample universe. The resting boxes were most positive in January at the Valparaíso

Reservoir and in December at the Bom Pastor Reservoir.

Natal & Marucci (1984) describe that the portable aspirator they developed can be used in Shannon traps, but this study proves that the same equipment can also be used in resting boxes.

This portable aspirator was also used in the study by Medeiros-Souza et al. (2013) collecting mosquitoes in parks in the city of São Paulo, where the Culex quinquefasciatus species predominated.

Sandhu et al. (2013) tested the resting box method in Riverside County, California, USA, with very different results to this study, but it is worth highlighting the methodological differences and local characteristics. Firstly, the objectives of that study were more related to entomological research into the risk of zoonotic transmission than the issue of nuisance.

The resting boxes used in this study were made of wood and had very different dimensions: they were 1.82 metres high, 1.82 metres deep and 1.2 metres wide (6 feet high, 6 feet deep and 4 feet wide), just big enough for a person to get inside, which was done in order to carry out the vacuuming with a portable hoover. This study was carried out on three farms, all of which contained horses and dogs, one of which also had a farm with more than 300 birds. All the properties bordered each other or had water sources inside, which could be potential breeding grounds for mosquitoes. There were fewer goats and employees. The resting boxes were installed close to the animals. Because most of the hosts were not human, being in a rural area, which could result in a greater variability of species, 98.3 per cent of the mosquitoes collected were of the Culex genus.

This North American study also used traps for pregnant females to compare with the resting boxes. Engorged females accounted for 70 per cent of all females collected and most of these engorged females were collected in the resting boxes. It is worth emphasising the difference between the two studies because the environment has a decisive influence. Our study was carried out in an urban area and the food source for the females is indoors, meaning that in order to achieve a slightly more similar result, the resting box would have to be installed inside the home, whereas in California it was carried out in a rural area where the food sources, which are the animals, are outdoors. Even so, the number of engorged females would probably be nowhere near the numbers in that study because, in addition to the physical barriers of human dwellings, the human population defends itself using physical, chemical and biological means, and the shape of the human body and its topographical anatomy allow for a better defence reaction to bites than horses, dogs and chickens. Climatological variables must also be taken into account.

Sandhu et al. (2013) showed that the Culex quinquefasciatus species accounted for 70.7 per cent of the females collected in the resting boxes, while in the pregnant traps this percentage reached 97.4 per cent. The resting boxes collected 28.8 times more females of the Culex genus and 32.2 times more females of the Culex quinquefasciatus species than the pregnant traps.

After the blood meal, the females try to rest in cool, dark places to digest and produce eggs. In the Riverside, California study, care was taken to install the resting boxes with the opening facing south to avoid sunlight.

In this study, this concern did not exist because the resting boxes were installed in shaded places, either by vegetation or protected by the shade of buildings or urban facilities such as pinguelas. Furthermore, Santo André has moderate sunlight for most of the year, and the resting box and consequently its opening was much smaller than in the Riverside study.

In Riverside, aspiration was carried out inside the resting boxes during the early hours of the morning until 9am at the latest, in favour of the idea that collection should be carried out during the coolest hours of the day. The authors of that work considered the resting box method for collecting engorged females to be a great success. They praised the method because it doesn't require attractants such as dry ice or batteries (the portable hoover must be battery-operated), it takes less than an hour to operate and it is not prone to vandalism or theft and is an ideal method, according to them, for developing countries.

If we judge for a rural environment, it may well be, but if we extrapolate to our urban reality, there is a total divergence.

Firstly, the litter box is fixed, it is difficult to transport, you can't carry it by hand, it gets in the way of passers-by and vehicles, it is prone to vandalism and can be mistaken for a public toilet or a hiding place for illegal acts, not to mention the fact that the population density in the city is much higher than in the countryside. In addition, in our work, the box is easy to transport and can be closed and vacuumed at any time of the day as long as it is not too long after collection, in a safe environment such as a laboratory. Another unfeasible factor is the proposed working hours, which can be up to 9am, but if it's in the early hours of the morning it will generate overtime and more costs for the management bodies. It's not clear why the resting box method generates fewer man-hours than vacuuming shelters, but we found that it takes twice as long and twice as many vehicles to carry out this method as the others. As for theft, it seems more difficult because of its size, but in our reality theft of recyclable materials such as wood is very common. During the course of our study, at least five boxes were stolen and then replaced, and they were made of material with no commercial value such as Styrofoam, painted and with the lid pierced. We also have to consider the huge socio-economic and cultural differences between cattle farmers from the richest state on the planet and Santo André, an industrial city suffering from the exodus of large companies and the majority of whose population comes from the proletariat, and these characteristics will certainly influence the dissemination of scientific production data.

In order to study the cycle of arboviruses and mosquito-borne diseases, it is important to capture engorged females after blood repast in order to find out which are the hosts and what they feed on. In a study carried out in Indonesia, Papua New Guinea and the Solomon Islands, Burkot et al. (2013) used protective screens stretched over bamboo poles to collect mosquitoes of the Culex quinquefasciatus species and applied the human blood index, which is the proportion of mosquitoes with human blood in their digestive tract over the total number of mosquitoes collected, because this allows them to assess the zoonotic potential of this species, despite its marked endophilia. In this study, they also use the resting box, but warn against collecting too few engorged females because, according to these authors, the resting box is an artificial container that is not very attractive compared to the natural shelters offered by the lush vegetation in these countries. Govella et al. (2011) share the same opinion.

Pombi et al. (2014) used an adhesive resting box and compared it with other types of traps in Burkina Faso and concluded that it is a good method, but made reservations about its use due to the peculiar geographical characteristics of each region. He emphasises that it is an inexpensive method.

Aspects of the seasonality related to the proliferation of Culex quinquefasciatus in the pools and determining the best time for control interventions were evaluated.

Cardoso (2010), carrying out entomological surveillance using the Nasci aspirator to collect adults in Rio Grande do Sul, concluded that the best periods for surveying and monitoring the culicidae fauna in that state are between October and May, which coincides almost entirely with the data from this study.

Seasonal aspects for the control intervention were also checked, using the medians of the indices, and it was found that the control of immatures in the aquatic phase is more effective and that August proved to be the month of choice for the general intervention to control mosquitoes.

Rubio et al. (2011) in the Buenos Aires region used a container index and found that tyres in shaded areas had a higher value than tyres exposed to sunlight for both Culex and Aedes. They also used a building index or site index and found that both indices for the two genera studied decreased as the degree of urbanisation increased and that they also increased in summer, and the building and container indices remained high for Culex even in autumn, especially in smaller towns.

Fischer et al. (2013), in the Netherlands, adopt indices (mathematical models and Floquet percentages) for the Culex quinquefasciatus mosquito, the vector of River Rift Fever, which despite being considered a zoonosis, like Foot and Mouth Disease, is a severe disease only in animals and is almost asymptomatic in humans. However, this disease, which is confined to the African continent and the Middle East, receives special attention in order to prevent its entry into Dutch territory, and risk maps are even used associating the presence of the possible vector with areas where cattle are reared. They establish that the vector season begins on 21 April and lasts until 23 October.

In our study, the population's annoyance indices in relation to mosquito bites around swimming pools were obtained and validated, but it is necessary to adjust the radius of the complaints specifically referring to Valparaíso because it is likely that the epicentre of the complaints is not the aforementioned swimming pool, due to the results of the samples found.

De La Mora-Covarrubias et al. (2008) in a study carried out in the cities of Ciudad Juarez, Mexico and El Paso, United States, proposed a 2 km radius for mosquito flight dispersal and analysed the data using statistical indices, spatial autocorrelation and dispersion maps, finding no significant differences in the 2006 data compared to 2005 and maintaining the same population density. CDC traps with dry ice as an attractant were used and only female mosquitoes were counted, as the aim was to assess the risk of arboviruses. Fourteen sampling points were used, 4 kilometres apart, thus avoiding overlapping spatial dispersion. Studies on the ecology of the vector should continue with the aim of determining the environmental, socio-demographic and physical factors that determine its environmental fluctuation over time. In this way, combining efforts to add geospatial measures of prediction with prognosis of abundance will make the control of Culex quinquefasciatus more efficient.

There is disagreement in the literature as to the ideal dispersal radius for Culex. Ryan et al. (2004), using a dispersal map using the Kriging technique, arrived at an ideal dispersal radius of 1500 metres for Culex annulirostris in Australia, while Turell et al. (2005) concluded that the mosquito's radius is 2000 metres.

Allen & Lu (2003) in a study carried out in Fairfax, Virginia, USA, suggested that the use of geostatistics associated with remote sensing systems was effective for mosquito surveillance and control. The researchers demonstrated the advantage of geospatial techniques such as autocorrelation indices and Kriging interpolation, which made it possible to estimate the abundance of the Culex *quinquefasciatus* mosquito, *the* boundaries of the sampling sites and the temporal frequency of sampling.

Jacob et al. (2008) carried out a study in a refugee camp in Gulu, Uganda in which satellite images captured by GPS were used to georeference the locations of Anopheles gambiae larvae based on shady places and Culex quinquefasciatus larvae were associated with aquatic vegetation. They applied statistical tests concluding that the Anopheles gambiae population forms significant clusters of population density and the same does not occur with the Culex quinquefasciatus population. The abundance of mosquito larvae can be measured using dynamic hydrological models. These models represent topographic variability and its control on soil moisture heterogeneity and runoff being measured remotely. A digital elevation model (DEM) was used to measure the hydrological variations of the soil, including the level of saturation and runoff, as well as to measure the existence of groundwater that may rise and the use of the land by the population and analyse whether the practices are conducive to mosquito proliferation. Refugee camps are susceptible to mosquito attack because of overcrowding with restricted access to water and inadequate water storage, creating mosquito breeding grounds.

Stephens et al. (1995) found in a study carried out in Tanzania that the population of villages is more sceptical and distrustful of health authorities, educators and malaria control staff when they suffer bites and nuisances related to Culex quinquefasciatus in their homes, making no distinction between the malaria vector and the nuisance factor.

Medlock and Vaux (2014) analyse "wetlands" built in urban areas in England which, despite bringing a sense of improved quality of life, are also a nuisance for the population because of mosquitoes.

LaDeau et al. (2013) in interviews with residents of the cities of Baltimore and Washington D.C. (USA) found that the rate of discomfort from mosquito bites reported was not related to the risk of acquiring diseases and the authors concluded that there should be motivation and clarification campaigns among the residents of these cities. They also found a higher number of immature mosquitoes in properties with lower purchasing power and incriminated discarded tyres as the most frequent breeding ground for the *Aedes* and *Culex* genera.

Cook et al. (2010), in a study in Scotland, questioned a number of residents about discomfort and mosquito bites to find out what factors cause people to be bitten or not and concluded that taller men and women with a higher body mass index are more likely to be bitten, but they state that as the data source was questionnaires, more studies are needed to reduce the subjectivity of the conclusions.

CHAPTER 7

CONCLUSIONS

This study was used to create four monitoring indices: the Density of Immatures per Litre (DIL), the Index of Mosquitoes Aspirated per Minute (IMAM), the Index of Mosquitoes Aspirated from Restrooms (IMACR) and the Index of Corresponding Complaints (IRC).

The indices proved to be easy to use, feasible and adapted to the reality of zoonosis or vector control services, both official and private. No more than two trained people are needed to apply the infestation indices.

All the indices proved to be valid, showing the properties of good indicators, and since the collection equipment is easy to acquire, they can be extrapolated to any work in the relevant scientific literature researched.

All of them have their perennial usefulness and the LID shows an even greater degree of standardisation of the revised larval indices.

The IMAM is universal and could be used to compare different levels of culicidae infestation or specific species for measuring the degree of adult mosquito infestation.

The IMACR is a valid index because it can currently be used in any mosquito-infested region, but its results will vary enormously due to the geographical, climatological, human and epidemiological characteristics of each region and the species of mosquito and its physiological state being studied.

As the IRC is an index based on complaints, it is more subjective and may therefore be less sensitive, but it should be improved with further studies of human behaviour, geographical references and statistical treatments.

This index was considered valid, especially in the case of Culex quinquefasciatus, which at certain times causes public control bodies to receive a veritable flood of complaints from the same region or neighbourhood, and this index will help find the main focus of the complaints through apps.

These indices were useful in determining the best time to control Culex quinquefasciatus in Santo André, but this data is very specific and will vary from region to region due to geographical characteristics and factors linked to public management.

CHAPTER 8

REFERENCES

Acha, PN; Szyfres, B. Zoonosis y enfermedades transmisibles comunes al hombre y los animales. Washington, 1989. 989 p.

Allen, TR; Lu, GY. Integrating remote sensing, terrain analysis, and geostatistics for mosquito surveillance and control. Proceedings of the American Society for Photogrammetry and Remote Sensing. 2003. Anchorage. 11 p.

Alves, SN; Tiburcio, JD; Melo, AL de. Susceptibility of Culex quinquefasciatus larvae to different insecticides. Journal of the Brazilian Society of Tropical Medicine. Uberaba vol. 44 n° 4 p.486-489. Jul / Aug. 2011.

Amonkar, SV; Banerj, A. Isolation and characterisation of larvicidal principle of garlic. Science News vol. 174 n° 4016 (dec 24, 1971) pp. 1343-1344.

Barata, EA et al. Capture of culicides in urban areas: evaluation of resting box method. Rev. Saúde Pública 2007 Jun, 41 (3): 375-382.

Barata, EAMF; Costa, AIP; Chiaravalotti Neto, F; Glasser, CM; Barata, JMS; Natal; D. Population of Aedes aegypti in a dengue endemic area, southeastern Brazil.Rev. Saúde Pública 2001; 35(3): 237-42.

Beer, R; Thomas, JA. Everything is water: With reservoirs at their lowest levels in history, Brazil is beginning to live with the frightening shadow of a shortage of the irreplaceable liquid, without which there is no life and the economy grinds to a halt. (Special) Veja, Oct 29, 2014, Vol.47(2397), p.84(4).

Bellen, HMV. Sustainability indicators: a comparative analysis. 2ª Ed. Rio de Janeiro. Editora FGV. 2006. 256 p.

Besen, GR; Gunther, WMR; Rodrigues, AC; Brasil, AL. Solid waste; vulnerabilities and perspectives. In Saldiva, P. Meio ambiente e Saúde; desafio das metrópoles. São Paulo, Instituto Saúde e Sustentabilidade. 2010.p. 106-123.

Biswas D, Mandal B, Biswas B, Banerjee A, Mukherjee TK. Plying of speedboats along canals in the city of Kolkata, India, to prevent mosquito breeding.Trans R Soc Trop Med Hyg. 2013 Mar; 107(3): 147-51

Bracco, JE. Evaluation of insecticide resistance in the Culex quinquefasciatus (**Diptera: Culicidae**) population of the Pinheiros River (São Paulo, Brazil). [Master's thesis - School of Public Health - USP] São Paulo, 1998.

Braks, MA; Cardé, RT. Improving efficacy of box gravid traps for collecting Culex quinquefasciatus. J Vector Ecol. 2007 Jun, 32 (1): 83-9. University of California.

Bueno - Mari, R; Correlia-López; E; Jimenez-Peydró; R. Culicidiofauna (**Diptera: Culicidae**) present in the different water enclaves of the city of Valencia (Spain). Revista Colombiana de Entomologia 36 (2). 2010.

Burkot, TR et al. Barrier screens: a method to sample blood-fed and host-seeking exophilic mosquitoes. Malaria Journal. 2013; 12: 49.

Bussab, Wilton. Basic Statistics. Saraiva. 5th edition 2006. 540p

Calisher, C H. Identification of a new Venezuelan equine encephalitis virus from Brazil. The American journal of tropical medicine and hygiene 1982 vol.:31 (6):1260 -72.

Cardoso, JC. Entomological surveillance of mosquitoes (**Diptera: Culicidae**) as a strategy for Environmental Health Surveillance in Rio Grande do Sul, Brazil. PhD thesis. School of Public Health - USP. 2010.

Cerf, DC; Georghiou, GP. Cross resistance to juvenile hormone analogues in insecticide-resistance strains of Musca domestica L. Pestic. Sci 1974; 5: 759-767.

Cezare, JP; Malheiros, TF; Philippi Jr., A. Avaliação da Política Ambiental e Sustentabilidade: Estudo de caso do Município de Santo André-SP. Environmental Sanitary Engineering. Vol. 12 n° 4 - Oct/Dec 2007, 417-425.

Chahad, S; Boff, MIC. Effect of black pepper extracts on Culex quinquefasciatus Say (**Diptera: Culicidae**) larvae. Entomological Society of Brazil. Annals Londrina. Entomological Society of Brazil, 1994; 23(1): 13-18.

Consoli, RAGB; Lourenço-de-Oliveira, R. Main mosquitoes of sanitary importance in Brazil. Fiocruz. Rio de Janeiro, 1994. 228 p.

Cook, J. I et al. (2010). To bite or not to bite! A questionnaire-based survey assessing why some people are bitten more than others by midges. BMC Public Health, 10, 275.

Cuba, LM; Situation of biological control in integrated mosquito management. [Master's thesis - School of Public Health-USP] São Paulo. 2004.

David, MR; Lourenço-de-Oliveira, R; Freitas, RM. Container productivity, daily survival rates and dispersal of Aedes aegypti mosquitoes in a high income dengue epidemic neighbourhood of Rio de Janeiro: presumed influence of differential urban structure on mosquito biology. Mem. Inst. Oswaldo Cruz 2009 Sep 104 (6): 927-32.

De La Mora-Covarrubias, A; Rubio-Arias, HO; Jimenez-Castro, JA. Entomological surveillance of Culex quinquefasciatus Say, 1823 vector of arboviral diseases in the urban area of Ciudad Juarez, Chihuahua, Mexico. Universidad y Ciencia, 24 (2) p. 101-109. 2008. Juarez Autonomous University of Tabasco. Villa Hermosa, Mexico.

De Meillon, B; Paing, M; Sebastian, A; Khan, ZH. Outdoor resting of Culex pipiens fatigans in Rangoon, Burma.Bull World Health Organ, 1967, 36 (1): 67-73.

De Paula, MB; Gomes, AC. Culicidae (**Diptera**) in an area under the influence of dam construction in the State of São Paulo. Revista de Saúde Pública vol. 41 n° 2 pág. 284-289 São Paulo 2007

Dias, JCP. Participation, decentralisation and control of endemic diseases in Brazil. In Barata, RB;

Briceno-Leon, R. Endemic diseases: Social, cultural and behavioural approaches. Editora Fiocruz. 376 p. Rio de Janeiro. 2000.

Dias, MAFS; Dias, J; Carvalho, LMV; Freitas, ED; Dias, PLS. Changes in extreme daily rainfall for São Paulo, Brazil. Springer Science+Business Media B.V. 2012. vol. 116, issue 3, pp. 705-722.

Domingos, MF. Aspects of the ecology of Aedes aegypti (Linnaeus) in Santos, São Paulo, Brazil. [PhD thesis, USP School of Public Health.] São Paulo, 2005.

Dos Anjos, AF; Navarro-Silva, MA. Culicidae (**Insecta: Diptera**) in an area of Atlantic Forest in the State of Paraná, Brazil. Acta Scientarum Biological Sciences (UEM), Jan. 2008 vol.30 (1) pág. 23-25.

Edman, J; Kittayapong, P; Linthicum, K; Scott, T. Attractant resting boxes for rapid collection and surveillance of Aedes aegypti inside houses. J Am Mosq Control Assoc. 1997 Mar, 13 (1): 24-7.

Environmental Protection Agency, 2008. Available at: http://www.epa.gov/heatislandeffect/mitigation/greenroofs.htm.

Figueiredo, LTM. Emerging arboviruses in Brazil. Revista da Sociedade Brasileira de Medicina Tropical 40 (2): 224-229 Mar - Apr, 2007.

Fischer, EAJ; Boender, GJ; Nodelijk, G; de Koeijer, AA; van Roermund, HJW. The transmission potential of Rift Valley fever virus among livestock in the Netherlands: a modelling study. Vet Res. 2013; 44(1): 58.

Fofana, D et. ali. Specific diversity and culicidian nuisance in the villages of N'gatty and Allaba in laguna area of Ivory Coast.Bull Soc. Pathol Exot. 2010 Dec; 103(5): 333-9.

Forattini, OP. Medical Culicidology: Identification, Biology, Epidemiology. São Paulo: Edusp; 2002. v. 2.

Forattini, OP. Medical Entomology. São Paulo: Edusp. 1962: v.1.

Freitas, CM; Carvalho, ML; Ximenes, EF; Arraes, EF; Gomes, JO. Socio-environmental vulnerability, disaster risk reduction and resilience building - lessons from the earthquake in Haiti and the heavy rains in the Serrana Region, Brazil Ciência & Saúde Coletiva [On-line] 2012, 17: [Consultation date: 5 / November / 2013] Available from: **<http://www.redalyc.org/articulo.oa?id=63023390022>IS** SN 1413-8123

Freitas, CM; Giatti, LL. Indicators of environmental sustainability and health in the Legal Amazon, Brazil. Cadernos de Saúde Pública, Rio de Janeiro, 25(6): 1251-1266 jun. 2009.

Freitas, RMC de; Faria, MA; Alves, SN; Mello, AL. Effects of ivermectin on Culex quinquefasciatus larvae. Revista do Instituto de Medicina Tropical. São Paulo, 38 (4): 293-297. 1996.

From the Centers for Disease Control and Prevention. Rapid assessment for vector borne diseases during the Midwest flood, United States, 1993. [No authors listed] JAMA, 1994 Jul 27; 272 (4): 264-5.

Garcia, CG; Escovar, JC; Londono, YB, Moncada, LA. Altitude and life tables of Culex quinquefasciatus (**Diptera: Culicidae**) populations. Revista Colombiana de Entomologia Bogotá vol. 36 (1) p. 62-66. 2010

Giroldo, J. Existing flood containment reservoirs in the RMSP: a contribution to analysing design, operation and maintenance. [Master's dissertation] Escola Politécnica da USP, São Paulo. 2003.

Gomes, AC; Forattini, OP. Culex mosquito shelters (**Diptera: Culicidae**) in rural areas. Rev. Saúde Pública 1990; 24 (5): 394-7.

Govella, NJ; Chaki, PP; Mpangile; JM; Killen, GF. Monitoring mosquitoes in urban Dar es Salaam: evaluation of resting boxes, window exit traps, CDC light traps, Ifakara tent traps and human landing catches. Parasit. Vectors, 2011. Mar 21, 4: 40. Dar es Salaam, Tanzania.

Grainge, M; Ahmed, S. Handbook of plants with pest control properties. 1988. 470 p.

Guanasekaran, K.; Prabakaran, G.; Balaraman, K. 2002. Efficacy of a floating, sustained release formulation of Bacillus thuringiensis sub israelensis in controlling Culex quinquefasciatus larvae in polluted water habitats. Acta Tropica, 83: 241-247.

Harbison, JE; Mathenge, EM; Misiani, GO; Mukabana; ER, Day; JF. A simple method for sampling indoor- resting malaria mosquitoes Anopheles gambiae and Anopheles funestus (**Diptera: Culicidae**) in Africa. J Med Entomol.2006 May; 43 (3):473-9.

Jacob, BG et. ali. Hydrological modelling of geophysical parameters of arboviral and protozoan disease vectors in Internally Displaced People camps in Gulu, Uganda. International Journal of Health Geographic. Mar, 2008 vol. (17) p. 11.

Jacobi, PR. São Paulo, an unsustainable metropolis - how can we overcome this reality? Cadernos Metropolitanos. São Paulo v. 13 n.29 pag. 219- 239 Jan/Jun 2013.

Jiang, Y; Mulla, MS. Laboratory and field evaluation of spinosad, a biorational natural product against larvae of Culex mosquitoes. Journal American Mosquito Control Association 2009, Dec. 25(4): 456-466.

Jones, JW; Weathersbee, AA; Efird, P; Meisch, MV. Evaluation of Bacillus sphaericus 2362 against Culex quinquefasciatus in septic dishes. J. Am. Mosq. Control. Assoc. 1990; 6(3): 496-499.

Karch, S; Asidi, N; Manzambi, ZM; Salaun, JJ. The culicidian fauna and its nuisance in Kinshasha (Zaire). Bull Soc Pathol Exot. 1993; 86(1):68-75.

Kirby, MJ; West, P; Green, C; Jasseh, M; Lindsay, SW. Risk factors for house-entry by culicine mosquitoes in a rural town and satellite villages in The Gambia. Parasit Vectors. 2008 Oct 21; 1(1): 41.

Kluck, M; Guimarães, JR, Ferreira, J; Promit, CA. Care quality management at Hospital das Clínicas de Porto Alegre: Implementation and validation of indicators. RAS_vol. 4 n°16 jul. - sep. 2002.

Komar, N; Pollack, RJ; Spielman; A. A nestable fibre pot for sampling resting mosquitoes. J Am

Mosq Control Assoc. 1995 Dec, 11 (4): 463-7.

Kweka; EJ; Mwang'onde, BJ; Kimaro, E; Msangi, S; Massenga, CP; Mahande, AM. A resting box for outdoor sampling of adult Anopheles arabiensis in rice irrigations schemes of lower Moshi, northern Tanzania. Malar. J. Apr 25, 8:82.

LaDeau, SL; Leisnham, PT; Biehler, D; Bodner, D. Higher Mosquito Production in Low-Income Neighborhoods of Baltimore and Washington, DC: Understanding Ecological Drivers and Mosquito-Borne Disease Risk in Temperate Cities Int J Environ Res Public Health. 2013 April; 10(4): 1505-1526.

Laporta, JL. Dengue and the infestation of Aedes aegypti in the municipality of Santo André. [PhD thesis presented at the School of Public Health - USP] São Paulo, 2004.

Laporta, G. Ecology of Culex quinquefasciatus and Culex nigripalpus in the Tietê Ecological Park, São Paulo, Brazil. [Master's dissertation in Public Health presented at the School of Public Health - USP] São Paulo, 2007.

Lopes, J.; Santos, F.P.; Zequi, J.A.C.; Petroni, D. 2010. Efficiency and Persistence of Three Commercial Products Based on Bacillus thuringiensis israelensis and Bacillus sphaericus in the Control of Culicidae (**Diptera**) in Effluent Treatment Lagoons. EntomoBrasilis, 3(3): 85-88.

Manda H; Arce, LM; Foggie, T; Shah, P; Grieco, JP; Achee, NL. Effects of irritant Chemicals on *Aedes aegypti* resting behaviour: is there a simple shift to untreated "safe sites"? PLoS Negl. Trop. Dis. 2011 Jul: 5 (7). Bethesda, Maryland, United States.

Margalit, J; Dean, D. The story of *Bacillus thunrigiensis* var. *israelensis* (Bti). J Am. Mosq. Control Assoc. 1985; 1 (1): 1-7.

Maricato, E; Ogura, AT; Comarú, F. Urban crisis, habitat production and disease. In Saldiva, P. Meio ambiente e Saúde; desafio das metrópoles. São Paulo, Instituto Saúde e Sustentabilidade. 2010. p. 46 - 65.

Medeiros-Sousa, AR; Ceretti Júnior, W; Urbinatti, PR; Natal, D; de Carvalho, GC; de Paula, MB; Fernandes, A; de Mello, MHSH, de Oliveira, RF; Orico, LD; Gonçalves, EFB; Marrelli, MT. Biodiversity of mosquitoes (**Diptera: Culicidae**) in São Paulo city parks I. Biota Neotropica vol.13 n° 1 pág. 317-321, mar. 2013 Campinas.

Medlock, J. M. and Vaux, A. G.C. (2014), Colonisation of a newly constructed urban wetland by mosquitoes in England: implications for nuisance and vector species. Journal of Vector Ecology, 39: 249-260.

Menezes, RMT. Bioecological aspects of *Culex quinquefasciatus* and *Ochlorathus scapularis* sheltered in habitats of the Tietê Ecological Park in the city of São Paulo, SP, Brazil. [PhD Thesis - USP School of Public Health.] São Paulo. 2002.

Mikulak E, Gliniewicz A, Królasik A, Sawicka B, Rabczenko D. Evaluation of effectiveness of several repellents against mosquito bites available at the Polish market.Przegl Epidemiol. 2012; 66(3): 479-85.

Ministry of the Environment. National Environment Council. Resolution no. 258 of 26 August 1999. Available at URL: http://www.mma.gov.br/port7conama/res/res99/res2599.ttml

Ministry of Health. Rapid Diagnosis of Municipalities for Entomological Surveillance of Aedes aegypti in Brazil. Methodology for evaluating the Breteau and Predial indices. Ministry of Health Manual. Brasília. 2005.62 p.

Ministry of Health. Health Surveillance Secretariat. Strategic Management Secretariat. Culex quinquefasciatus Surveillance and Management Guide. Brasília, 2011. 71 pages.

Montes, J. Culicidae fauna of Serra da Cantareira, São Paulo, Brazil. Revista de Saúde Pública 39 (4): 578-584. 2005.

Morais, SA de. Aspects of Culex quinquefasciatus (**Diptera: Culicidae**) infestation in the Pinheiros River, São Paulo (São Paulo, Brazil) and considerations on population control applications. São Paulo, 2005. [Master's dissertation - USP School of Public Health].

Morais, SA de; Marrelli, MT. Aspects of the distribution of Culex (Culex) quinquefasciatus Say (**Diptera, Culicidae**) in the region of the Pinheiros River, in the city of São Paulo, State of São Paulo, Brazil. Brazilian Journal of Entomology. São Paulo vol. 50 n° 3 pág. 413-418. 2006.

Morais, SAde; Natal, D. The Culex quinquefasciatus mosquito (**Diptera: Culicidae**) in a municipality crossed by a river with a high pollution load. Brazilian Journal of Environmental Sciences. São Paulo n° 6 pág. 21-26. 2007.

Morais, SA de; Marrelli; M.T.; Natal, D. Observations on control applications in ecotopes with high infestation of Culex quinquefasciatus Say (**Diptera, Culicidae**). Revista Brasileira de Entomologia, 51: 246-251. 2007.

Morrison, AC; Rosenberg,R; Scott,TW ; Zielinsky-Gutierrez,E. Defining challengers and proposing solutions for control of the virus vector Aedes aegypti. PLoS Medicine5.3 (Mar.2008): p362.

Murrel, EG; Juliano, SA. Predation resistance does not trade off with competitive ability in early-colonising mosquitoes. Oecologia, November 2013, vol. 173 Issue 3 pp. 10331042.

Nasci, RS. A light weight battery powered aspirator for collecting resting mosquitoes in the field. Mosq. News. 1981; 41:808-11.

Natal, D. Importância epidemiológica de Culex do Subgênero Melanoconion (**Diptera: Culicidae**) [Master's thesis - USP School of Public Health]. São Paulo, 1981.

Natal, D & Marucci, D. Suction apparatus for catching mosquitoes. Rev. Public Health, 18: 418-20.1984

Natal, D; Paganelli, CH; Santos, JLF. Composition of the adult population of Culex quinquefasciatus Say, 1823 in ecotopes near the Edgard de Souza Dam, in the municipality of Santana do Parnaíba, State of São Paulo, Brazil. Rev. Bras. Entomol., 35: 539-43, 1991.

Nobre, CA et al. Vulnerability of Brazilian megacities to climate change. São Paulo Metropolitan

Region - Executive Summary São Paulo INPE /Unicamp/USP/IPT/UNESP. Rio Claro. 2010. 186 p.

Okumu, FO; Madumla, EP; John, AN; Lwetoijera, DW; Sumaye, RD. Attracting, trapping and killing disease-transmitting mosquitoes using odour-baited stations - The Ifakara Odor Baited Stations. Parasit. Vectors 2010; 3:12.

Oliveira, CL et al. Dengue incidence related to climatic conditions in the municipality of Toledo, PR. Arq. Cienc. Saúde UNIPAR, Umuarama, vol. 11 n° 3, pág. 211-216, Sep. - Dec. 2007.

Perich, MJ; Clair, PM; Boobar, LR. Integrated use of planaria (Dugesia dorotocephala) and Bacillus thunrigiensis var. israelensis (Bti) against Aedes taeniorrhynchus: a laboratory bioassay. J. Am. Mosq. Control Assoc. 1990; 6 (4): 667-671.

Philippi Jr, A; Malheiros, TF, editors. Various authors. Sustainability indicators and environmental management. Barueri 2012. Editora Manole. 743 p.

Philippi Jr, A; Pelicioni, MCF. Editors. Various authors. Environmental education and sustainability. Barueri, 2005. Manole Publishing House. 1004 p.

Pizarro, APB; Oliveira Filho, AM; Parente; JP; Melo, MTV; dos Santos, E; Lima, PR. The use of sisal industry waste to combat mosquito larvae. Rev. Med. Trop. jan. - feb. 1999; 32(1): 23-29.

Pombi, M. et al. (2014). The Sticky Resting Box, a new tool for studying resting behaviour of Afrotropical malaria vectors. Parasites & Vectors, 7, 247.

Pontes, R.J.S.; Regazzi, A.C.F.; Lima, J.W.O.; Kerr-Pontes, L.R.S. 2005. Residual effect of commercial presentations of the larvicides temephos and Bacillus thuringiensis israelensis on Aedes aegypti larvae in containers with water renewal. Revista da Sociedade Brasileira de Medicina Tropical, 38: 316-321.

Rasgon, JL. Editor. Various authors. *Wolbachia* Induces Male-Specific Mortality in the Mosquito *Culex pipiens* (LIN Strain). PLoS ONE 7(3): e30381. doi: 10.1371/journal.pone.0030381 (2012).

Rodrigues, HC. Effect of aqueous extracts of *Meliaceae* on the development of *Spodoptera frugiperda* (JE Smith, 1797). [Master's dissertation - USP]. Piracicaba, 1995.

Rubio, A; Cardo, M; Vezzani, D. Tire-breeding mosquitoes of public health importance along an urbanisation gradient in Buenos Aires, Argentina. Memórias do Instituto Oswaldo Cruz, Vol. 106, No. 6, 2011, pp. 678-684

Rutkowski, E; Jacobi, PR; Souza, RMG; Tagnin; R. The metropolitanised waters of Alto Tietê. In Saldiva; P. et al. Environment and Health: the challenge of the metropolises. São Paulo. Instituto Saúde e Sustentabilidade. 2010. pp. 88 - 105.

Ryan, PA; Lyon, AS; Alsemgeest, D; Thomas, P; Kay, BH. Spatial statistical analysis of adult mosquito (**Diptera: Culicidae**) counts: An example using light trap data, in Redlandshire, Southeastern Queensland, Australia. Journal of Medical Entomology 41 (6): 1143-1156. 2004.

Sales, S; Hervy, J.P. Rapport sur l' evaluation de la remanencede de cinq composes insecticides (OMS 786, OMS 1170, OMS 1197, OMS 1211, OMS 1424) controls populations larvaires de *Aedes aegypti* vecteur domestique de fiévre jaune, dans lês conditions locales de stockage de l'eau. OCCGE. Centre Muray. 1974.

Sandhu, TS; Williams, GW; Dhillon, MS. Population dynamics of blood-fed female mosquitoes and comparative efficacy of resting boxes in collecting them from northwestern part of Riverside County, California. Journal Global Infectious Diseases, 5: 15-18. 2013.

Santos, LU; Souza, AB; Andrade, CFS; Souza, CEP. Use of *Bacillus thunrigiensis* (H-14) as a mosquito control agent in a cemetery. Rev. Pat. Trop. 1994, 23 (2): 151-158.

Silvério, EC. Study of the mosquito fauna (**Diptera: Culicidae**) in flood containment reservoirs in the metropolitan area of the city of São Paulo, SP. [Master's dissertation. School of Public Health, University of São Paulo]. São Paulo, 2008.

Silvério, EC; Urbinatti, PR. The presence of mosquitoes (**Diptera: Culicidae**) in flood containment reservoirs in the eastern region of the City of São Paulo, Brazil. Rev. Soc. Bras. Med. Trop. 2011 Jul-Aug, 44 (4): 504-7.

Stephens, C et. ali. Knowledge of mosquitoes in relation to public and domestic control activities in the cities of Dar es Salaam and Tanga. Bull World Health Organ. 1995; 73(1): 97-104.

Suguimoto, RS; Natal, D. Variations in the mosquito fauna (Diptera: Culicidae) in an area where a hydroelectric dam is being built on the Paranapanema River, SP/PR. Master's dissertation. School of Public Health. 2009.

Tauil, PL. Perspectives for the control of vector-borne diseases in Brazil. Revista da Sociedade Brasileira de Medicina Tropical 39 (3): 275-277 May-June, 2006.

Tsuda, Y et al. After effects of tsunami on distribution and abundance of mosquitoes in rice-field in Miyagi Prefecture, Japan in 2011. Medical Entomology and Zoology vol. 63 iss: 1 pag. 21-30.

Turrel, MJ; Dohm, DJ; Sardelis, MR; O'Guinn, ML; Andreadis, TG; Blow, JA. An update on the potential of North American mosquitoes (**Diptera: Culicidae**) to transmit West Nile virus. Journal of Medical Entomology 42: 57-62. 2005.

Urbinatti, PR. Mosquito immatures (**Diptera, Culicidea**): characterisation of breeding sites - Parque Ecológico do Tietê, Guarulhos, SP, Brazil. [Master's dissertation - USP School of Public Health]. São Paulo, 2000.

Van den Bijllaardt, R et ali. The suitability of clay pots for indoor sampling mosquitoes in an arid area in northern Tanzania. ActaTrop. 2009 Aug. 111(2): 197-9. Nijmegen, the Netherlands.

Vitor Neto, JB. Review of control methods and resistance mechanisms of Culex quinquefasciatus (**Diptera: Culicidae)** [Master's thesis - USP School of Public Health]. São Paulo, 2006.

Wilke, ABB; Gomes, AC; Natal, D; Marrelli, MT. Vector control using genetically modified mosquitoes. Review. Rev. Saúde Pública 2009; 43(5): 86974.

World Health Organisation- WHO. Chemical methods for the control of vectors and pests of public health importance. Division of Control of Tropical Diseases/ WHO Pesticides Evaluation Scheme. Geneva. 1997. 108 p.

Wright, MJ. The resistance of the larval mosquito to cold, notes on the habits and life: History of mosquitoes in Aberdeenshire. British Medical Journal [0007-1447] vol. 1 iss: 2012 page: 882-883. 1901.

www.santoandre.sp.gov.br/index.php/noticias. Consulted on 14/12/2013. PMSA, 2013

www.saude.gov.br/portaldasaude/inseticida/controle of vectors. Consulted on 12/09/2013.

Yebakima, A; Raymond, M; Marquine, M & Pasteur, N. Resistance to organophosphurous insecticides in Culex quinquefasciatus (**Diptera: Culicidae**) from Martinique. J. Med. Entomol., 32 (2): 77-82, 1995.

Zhang, L; Harada, K; Shono, T. Cross resistance to insect growth regulators in pyriproxyfen-resistance housefly. Apply Entomology. Zool. 1998; 33: 195-197.

CHAPTER 9

APPENDICES

9.1 APPENDIX A - COLLECTION OF IMMATURE AND ADULT MOSQUITO SPECIMENS IN THE SANTO ANDRÉ FLOOD CONTAINMENT RESERVOIRS - PHOTOGRAPHIC ILLUSTRATIONS.

PHOTO 1 - VIEW OF THE LARVAE COLLECTION POINT IN THE RM-9 POOL, FACULTY OF MEDICINE (VALPARAÍSO), NEAR THE SPILLWAY, REFERRED TO AS COLLECTION POINT LV-1.

PHOTO 2 - GENERAL VIEW OF THE LARVAE COLLECTION POINT IN THE RM-9 POOL, FACULTY OF MEDICINE (VALPARAÍSO), NEAR THE PUMPS, COLLECTION POINT LABELLED LV-2.

PHOTO 3 - VIEW OF THE LARVAE COLLECTION POINT AT THE RM-9 SWIMMING POOL, FACULTY OF MEDICINE (VALPARAÍSO), NEAR THE PUMPS, COLLECTION POINT LABELLED LV-2.

PHOTO 4 - GENERAL VIEW OF THE LARVAE COLLECTION POINT IN POOL RM-9, FACULTY OF MEDICINE (VALPARAÍSO), NEAR THE BIRD NESTING SITE, LABELLED LV- 3. IN THE BACKGROUND, THE NORTH-WESTERN EDGE OF THE POOL.

PHOTO 5 - GENERAL VIEW OF THE LARVAE COLLECTION POINT IN THE RM-9 POOL, FACULTY OF MEDICINE (VALPARAÍSO), LABELLED COLLECTION POINT LV-4.

PHOTO 6 - GENERAL VIEW OF THE LARVAE COLLECTION POINT IN THE RM-9 POOL, FACULTY OF MEDICINE (VALPARAÍSO), LABELLED LV-4, AFTER CLEANING, SHOWING THAT IT IS A CHANNEL.

PHOTO 7 - LARVAE COLLECTION POINT AT THE RM-9 SWIMMING POOL, FACULTY OF MEDICINE (VALPARAISO), LABELLED LV-5.

PHOTO 8 - LARVAE COLLECTION POINT AT THE RM-9 SWIMMING POOL, FACULTY OF MEDICINE (VALPARAISO), LABELLED LV-5 AFTER CLEANING.

PHOTO 9 - VIEW OF THE ADULT MOSQUITO COLLECTION POINT AT THE RM-9 MEDICAL SCHOOL (VALPARAISO) SWIMMING POOL, LABELLED AV-1.

PHOTO 10 - VIEW OF THE ADULT MOSQUITO COLLECTION POINT AT THE RM-9 MEDICAL SCHOOL (VALPARAISO) SWIMMING POOL LABELLED AV- 2.

PHOTO 11 - VIEW OF THE ADULT MOSQUITO SUCTION POINT AT THE RM-9 SWIMMING POOL, FACULTY OF MEDICINE (VALPARAISO), REFERRED TO AS COLLECTION POINT AV-3. PHOTO TAKEN ON 6 AUGUST 2014

PHOTO 12 - VIEW OF THE ADULT MOSQUITO SUCTION POINT AT THE RM-9 MEDICAL SCHOOL (VALPARAÍSO) SWIMMING POOL, CALLED THE COLLECTION POINT AV- 3. WEEDING WORK BEING CARRIED OUT SIMULTANEOUSLY ON 23 SEPTEMBER 2014.

PHOTO 13 - VIEW OF THE ADULT MOSQUITO ASPIRATION POINT RM-9 MEDICAL SCHOOL (VALPARAÍSO) CALLED THE COLLECTION POINT AV- 4. IN USE, THE NASCI ASPIRATOR.

PHOTO 14 - GENERAL VIEW OF THE ADULT MOSQUITO ASPIRATION POINT AT THE RM-9 MEDICAL SCHOOL (VALPARAISO), LABELLED AV-5.

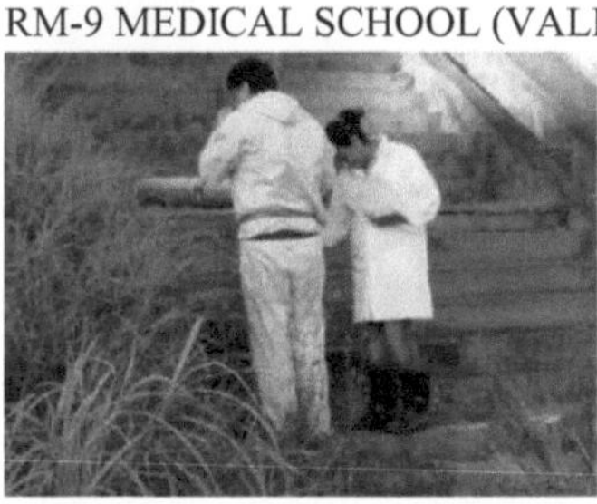

PHOTO 15 - VIEW OF THE ADULT MOSQUITO COLLECTION POINT AT THE RM-9 MEDICAL SCHOOL (VALPARAISO) SWIMMING POOL, LABELLED AV-5.

PHOTO 16 - VIEW OF THE RESTING BOX AT THE RM-9 MEDICAL SCHOOL (VALPARAISO) SWIMMING POOL, KNOWN AS CR-1.

PHOTO 17- VIEW OF THE RESTING BOX AT THE RM-9 MEDICAL SCHOOL (VALPARAISO) SWIMMING POOL, KNOWN AS CR-2.

PHOTO 18 - VIEW OF THE RESTING BOX AT THE RM-9 MEDICAL SCHOOL (VALPARAISO) SWIMMING POOL, KNOWN AS CR-3.

PHOTO 19 - VIEW OF THE RESTING BOX AT THE RM-9 MEDICAL SCHOOL (VALPARAISO) SWIMMING POOL, KNOWN AS CR-4.

PHOTO 20 - VIEW OF THE RESTING BOX AT THE RM-9 MEDICAL SCHOOL (VALPARAISO) SWIMMING POOL, KNOWN AS CR-5.

PHOTO 21 - VIEW OF THE RETENTION RESERVOIR IN JARDIM BOM PASTOR, THE IMMATURE COLLECTION POINT IS IN THE BOTTOM RIGHT-HAND CORNER OF THE IMAGE AND IS CALLED LBP-1.

PHOTO 22 - SITE DESIGNATED AS LBP-1 FOR COLLECTING IMMATURES

PHOTO 23 - ATTEMPT TO COLLECT IMMATURES AT POINT LBP-1. HIGHLIGHTED, THE ENTOMOLOGICAL SHELL.

PHOTO 24 - PROCEDURE FOR COLLECTING IMMATURES AT POINT LBP-2.

PHOTO 25 - SITE DESIGNATED AS LBP-3 FOR COLLECTING IMMATURES. RUA FELÍCIO PEDROSO IS TO THE BACK OF THE PHOTOGRAPHER.

PHOTO 26 - SITE DESIGNATED AS LBP-4 FOR COLLECTING IMMATURES.

PHOTO 27 - SITE DESIGNATED AS LBP-5 FOR COLLECTING IMMATURES.

PHOTO 28 - VIEW OF THE SURROUNDINGS OF THE JARDIM BOM PASTOR RETENTION RESERVOIR, THE ADULT COLLECTION POINT IS CALLED ABP-1.

PHOTO 29 - VIEW OF THE SURROUNDINGS OF THE JARDIM BOM PASTOR RETENTION RESERVOIR, THE ADULT COLLECTION POINT IS CALLED ABP-2.

PHOTO 30 - VIEW OF THE RETENTION RESERVOIR AT JARDIM BOM PASTOR, THE ADULT COLLECTION POINT IS CALLED ABP-3.

PHOTO 31 - SAMPLING OF THE ABP-3 ASPIRATION COLLECTION POINT, WHERE MORE THAN 600 CULICIDAE WERE COLLECTED ON 18 DECEMBER 2014.

PHOTO 32 - VIEW OF THE RETENTION RESERVOIR IN JARDIM BOM PASTOR, THE ADULT COLLECTION POINT IS CALLED ABP-4. YOU CAN SEE THE AFOREMENTIONED PINGUELA IN THE BACKGROUND.

PHOTO 33 - VIEW OF THE RETENTION RESERVOIR AT JARDIM BOM PASTOR, THE ADULT COLLECTION POINT IS CALLED ABP-5.

PHOTO 34 - VIEW OF THE RETENTION RESERVOIR AT JARDIM BOM PASTOR, THE ADULT COLLECTION POINT IS CALLED ABP-5 AND THE SAMPLE WAS TAKEN ON 18 DECEMBER 2014.

PHOTO 35 - VIEW OF THE RESTING BOX OF THE RETENTION RESERVOIR IN JARDIM BOM PASTOR, CALLED CR-6.

PHOTO 36 - VIEW OF THE RESTING BOX OF THE RETENTION RESERVOIR IN THE BOM PASTOR GARDEN, CALLED CR-7. IN THE BACKGROUND THE FENCE OF THE GARDENING COMPANY.

PHOTO 37 - VIEW OF THE RESTING BOX OF THE RETENTION RESERVOIR IN JARDIM BOM PASTOR, KNOWN AS CR-8.

PHOTO 38 - VIEW OF THE RESTING BOX OF THE RETENTION RESERVOIR IN JARDIM BOM PASTOR, KNOWN AS CR-9.

PHOTO 39 - VIEW OF THE RESTING BOX OF THE RETENTION RESERVOIR IN JARDIM BOM PASTOR, KNOWN AS CR-10.

PHOTO 40 - VIEW OF THE RESTING BOXES BEING PREPARED.

PHOTO 41 - USING THE HOOVER.

PHOTO 42 - GENERAL VIEW OF THE FLOODED RM-9 MEDICAL SCHOOL POOL SPILLWAY ON 27 JANUARY 2015. COMPLETE DRAINING TOOK PLACE IN THE AFTERNOON OF THE SAME DAY.

PHOTO 43 - CHRISTMAS AND MARUCCI HOOVER DISASSEMBLED WITH 12 V BATTERY PACKED IN HOLSTER BELT AND COLLECTION CUPS AND COVER. IN THIS WORK A 6 V BATTERY WAS USED WHICH WAS CONNECTED TO A CHARGER WHICH THEN CHARGED THE HOOVER.

9.2 APPENDIX B - SCHEDULE OF WEEDING CARRIED OUT AT THE MEDICAL SCHOOL RESERVOIR (VALPARAÍSO) AND BOM PASTOR RESERVOIR

Chart 8 - Weeding schedule carried out by PMSA, the Government of São Paulo and private individuals at the RM-09 Reservoirs at the Faculty of Medicine and the Bom Pastor Retention Reservoir between August 2014 and May 2015.

Execution	Valparaíso Swimming Pool	Bom Pastor Swimming Pool
State	September 2014	
Municipal		September 2014
- Private		October 2014
Municipal		November 2014
Municipal		January/2015
Municipal		March 2015
State	March 2015	
Municipal		May/2015

Source: Drainage Management/DMO/ SEMASA /PMSA and DAEE- São Paulo State Government.

- Weeding was carried out by an employee of a landscaping company in a partial area next to the lagoon on Rua Xingu, which belongs to the Bom Pastor pool.

Photo 1 of Appendix B - Photograph of weeding work at the RM-9 Faculdade de Medicina Reservoir (Valparaíso) taken on 23/09/2014. According to the DAEE source, an area of 7920 m^2 was weeded.

Photo 2 in Appendix B - Photograph of a weeding service carried out by an employee of a landscaping company at the Bom Pastor Reservoir, taken on 30/10/2014.

Photos 3 and 4 in Appendix B - Photographs showing the weeding that took place in the Bom Pastor Reservoir area, taken on 30 October and 18 December 2014 respectively. Weeding was carried out in November by the municipal team from the Drainage Management/ DMO/ SEMASA/PMSA.

9.3 APPENDIX C - SCHEDULE OF MECHANICAL CLEANING CARRIED OUT AT THE MEDICAL SCHOOL RESERVOIR (VALPARAÍSO) AND BOM PASTOR RESERVOIR.

Table 9- Mechanical Cleaning Schedules carried out on the RM- 09 Reservoirs at the Faculty of Medicine and the Bom Pastor Retention Reservoir between August 2014 and May 2015.

Execution	**Valparaíso Swimming Pool**	**Swimming poolGo od Pastor**
Municipal		**August 2014**
State	**September 2014**	
Municipal		**January 2015**
State	**January 2015**	
State	**February 2015**	

Source: Drainage Management/DMO/SEMASA/PMSA and DAEE/São Paulo State Government.

Note: During the cleaning of the Valparaíso pool in January 2015, 1675 $m.^3$ of mud and sludge were removed by excavation, according to information from the DAEE.

Photo 1 in Appendix C - Excavator used for mechanical cleaning in the Valparaíso pool. Photo taken on 17/09/2014.

Photo 2 in Appendix C - Mechanical cleaning being carried out in the Valparaíso pool. Photo taken on 21/01/2015.

Photo 3 in Appendix C - Mechanical cleaning being carried out in the Bom Pastor pool with a floating excavator, known as a Pantaneira. Source: Drainage Management/DMO/SEMASA/PMSA.

Photo 4 in Appendix C - Mechanical cleaning being carried out in the Bom Pastor pool with a floating excavator, known as a Pantaneira. Source: Drainage Management/DMO/SEMASA/PMSA.

Photo 5 in Appendix C - Mechanical cleaning being carried out in the Bom Pastor pool with a floating excavator, known as a Pantaneira. Source: Drainage Management/DMO/SEMASA/PMSA.Moving earth to the banks of the stream 13/08/2014 - 13m^3 09/01/2015 - 50m^3

Source: Drainage Management/DMO/SEMASA/PMSA

APPENDIX D - SCHEDULE OF CHEMICAL AND BIOLOGICAL TREATMENT CARRIED OUT AT THE MEDICAL SCHOOL RESERVOIR (VALPARAÍSO) AND BOM PASTOR RESERVOIR.

Table 10 - List of biocidal products with active ingredient and trade name with respective dosages and concentrations used in the RM-09 Reservoirs at the Faculty of Medicine and the Bom Pastor Retention Reservoir between August 2014 and May 2015.

DATE	LOCAL	ACTIVE INGREDIENT	PRODUCT USED	PRESENTATION	*CONC.	DOSAGE	VOLUME USED
12/11/14	SURROUNDINGS OF THE XINGU STREET LAGOON (PISCI-NÃO BOM PASTOR)	*BACILLUS SPHAERI-CUS* CRYSTALS AND SPORES	VECTOLEX CG	LARVICIDE BIOLOGICAL IN BASED GRANULES CORN COB POWER: 670 UTI /MG	7,5 %	THE LAUNCH	3 KG
24/11/14	IGUAPE AND FELÍ-CIO PE-DROSO STREETS (PISCI-NO BOM	*BACILLUS SPHAERI-CUS* CRYSTALS AND SPORES	VECTOLEX CG	LARVICIDE BIOLOGICAL IN BASE GRANULES CORN COB POWER: 670 UTI /MG	7,5 %	THE LAUNCH	4 KG

	PASTOR)						
15/12/14	WINTER STREET, AVENUE SARINA, GUA-TAM-BÚS AND GUA-DARAMA STREETS	PIRIMIFÓS METALLIC	ACTELLIC PROF	EMULSIFIABLE CONCENTRATED ORGANOPHOSPHATE INSECTICIDE	50%	30 ML/ 10 I	100 LITRES OF SYRUP
21/12/14	SURROUNDINGS OF THE XINGU STREET LAGOON (PISCI-NÃO BOM PASTOR)	PYRIMIFOS METHYL	ACTELLIC PROF	EMULSIFIABLE CONCENTRATED ORGANOPHOSPHATE INSECTICIDE	50%	67.5 ML/10 1	100 LITRES OF SYRUP
08/04/15	SURROUNDINGS OF THE XINGU STREET LAGOON (PISCI-NÃO BOM PASTOR)	*BACILLUS THUNRIGI- ENSIS ISRAELEN- SIS* (BTI) CRYSTALS AND SPORES	TEKNAR	BIOLOGICAL LARVICIDE IN SUSPENSION	1,6%	300 ML/10 I	100 LITRES OF SYRUP
21/04/15	PISCI-NO VALPA-	*BACILLUS THUNRIGI- ENSIS ISRAELEN*	TEKNAR	LARVICIDE BIOLOGICAL IN SUSPENSION	1,6%	120 ML/ 10 1	100 LITRES OF SYRUP

	RAÍSO	- *SIS* (BTI) CRYSTALS AND SPORES						

SOURCE: ECRVAS/GCZ/DVS/SMS/PMSA *CONC.= PRODUCT CONCENTRATION

Approximate area of the Xingu Street Lagoon: 28,000 m.^{2}

MANUFACTURER'S RECOMMENDATIONS:

- **VECTOLEX CG**

MANUFACTURER: ECOSYSTEM

DOSAGE: Polluted waters - 5.6 to 11.2 Kg per hectare

Deep water or to maintain residual: - 11.2 to 22.5 kg per hectare

- **ACTELLIC PROF 50%**

MANUFACTURER: SYNGENTA

DOSAGE: 10 to 100 millimetres per 10 litres of water for spraying 1000 square metres.

- **TEKNAR 1,6 %**

MANUFACTURER: SYNGENTA

DOSAGE: 200 millimetres per 10 litres of water for spraying 300 square metres.

SOURCES CONSULTED:

www.semasa.sp.gov.br/drenagem/macrodrenagem/piscinões

www.syntinela.com.br/website/ media/pdf/products-and-brands/pest-control-urban-and-garden/products/actellic-prof/actellicprof.pdf

www.syntinela.com.br/website/ media/pdf/products-and-brands/pest-control-urban-and-garden/products/teknar/teknar.pdf

www.ecosystem.com.br/#vectolex-cg/c18a7

9.5 APPENDIX E - APPENDIX OF DAILY RAINFALL MEASURED IN THE REGION FOR THE MEDICAL SCHOOL (VALPARAÍSO) AND BOM PASTOR RESERVOIRS.

Table 11 - Values measured by the Remote Transmission Unit (RTU-78) for daily rainfall in the region comprising the RM-09 Reservoirs at the Faculty of Medicine and the Bom Pastor Retention Reservoir between August 2014 and May 2015.

Table -	Daily rainfall during the months in which fieldwork took place.																			
Aug/14			Sep/14		Oct/14		Nov/14		Dec/14		Jan/15		Feb/15		Mar/15		Apr/15		May/15	
Day		Prec.	Day	Prec.	Day	Prec.	Day	Prec.	Day	Prec.	Day	Prec.	Day	Prec.	Day	Prec.	Day	Prec.	Day	Prec.

01		0,0	01	0,0	01	0,4	01	0,2	01	0,0	01	0,0	01	0,0	01	2,2	01	0,0	01	0,0
02		0,0	02	12,7	02	3,0	02	0,0	02	0,0	02	13,8	02	52,0	02	0,0	02	0,4	02	0,0
03		0,0	03	5,8	03	0,4	03	0,4	03	0,0	03	1,8	03	0,0	03	0,0	03	0,0	03	0,0
04		0,0	04	2,6	04	0,0	04	10,2	04	6,8	04	3,6	04	4,2	04	0,0	04	0,0	04	3,2
05		3,2	05	0,0	05	0,0	05	2,2	05	0,8	05	2,2	05	39,2	05	0,0	05	0,2	05	0,4
06		0,0	06	0,0	06	0,0	06	0,4	06	0,2	06	0,2	06	17,9	06	19,1	06	1,0	06	0,0
07		0,0	07	0,0	07	0,0	07	0,8	07	0,0	07	1,2	07	0,6	07	59,4	07	11,4	07	2,8
08		1,4	08	0,0	08	0,0	08	5,4	08	0,0	08	3,8	08	0,0	08	19,4	08	0,0	08	0,2
09		0,0	09	0,0	09	0,0	09	0,0	09	0,4	09	0,0	09	1,8	09	4,6	09	0,0	09	0,0
10		0,0	10	0,0	10	0,0	10	0,0	10	4,0	10	2,8	10	4,6	10	1,6	10	0,0	10	56,1
11		0,0	11	0,0	11	0,0	11	0,0	11	11,6	11	4,8	11	14,3	11	26,5	11	0,0	11	0,0
12		0,0	12	0,0	12	0,0	12	0,0	12	30,3	12	9,6	12	0,0	12	2,6	12	0,0	12	6,4
13		11,4	13	0,0	13	0,0	13	6,2	13	6,0	13	0,0	13	0,8	13	2,4	13	0,0	13	0,8
14		0,4	14	0,0	14	0,0	14	2,0	14	9,8	14	22,2	14	0,0	14	1,8	14	0,0	14	0,0
15		0,0	15	0,6	15	0,0	15	1,0	15	0,0	15	0,0	15	15,7	15	9,8	15	23,5	15	0,0
16		3,0	16	0,0	16	0,0	16	0,2	16	0,0	16	0,0	16	13,6	16	4,2	16	0,0	16	0,0
17		0,2	17	0,0	17	0,0	17	0,0	17	0,0	17	0,6	17	15,7	17	0,2	17	0,0	17	0,0
18		0,0	18	0,0	18	0,0	18	0,0	18	0,0	18	0,0	18	9,8	18	0,0	18	0,0	18	0,0
19		0,0	19	4,6	19	0,2	19	0,0	19	3,0	19	0,0	19	0,0	19	5,8	19	0,4	19	0,0
20		2,6	20	2,6	20	1,2	20	0,0	20	0,2	20	0,0	20	0,0	20	12,2	20	0,0	20	0,0
21		0,0	21	7,2	21	1,6	21	0,0	21	0,0	21	8,0	21	0,0	21	5,4	21	3,4	21	0,0
22		0,0	22	0,0	22	0,0	22	0,0	22	49,5	22	31,2	22	0,0	22	0,4	22	11,8	22	0,0
23		0,0	23	0,0	23	0,0	23	2,0	23	14,4	23	2,0	23	0,0	23	0,0	23	0,0	23	0,0
24		0,0	24	0,0	24	0,0	24	0,0	24	1,4	24	25,5	24	13,2	24	0,0	24	0,0	24	0,0
25		0,0	25	0,4	25	0,0	25	23,1	25	1,4	25	10,6	25	20,8	25	0,0	25	1,0	25	0,0
26		6,0	26	29,	2	1,8	2	34,9	26	0,0	2	37,	26	0,8	26	0,0	26	0,2	2	0,

				1	6		6				6	0							6	0
27		1,2	27	0,2	27	0,0	27	0,6	27	0,0	27	47,2	27	0,4	27	0,0	27	0,0	27	0,4
28		0,6	28	0,6	28	0,0	28	0,0	28	0,0	28	6,8	28	39,5	28	0,0	28	0,0	28	4,4
29		0,0	29	0,2	29	0,0	29	0,0	29	9,2	29	0,2			29	9,8	29	0,4	29	3,6
30		0,0	30	0,0	30	0,0	30	0,0	30	0,0	30	2,0			30	0,4	30	0,2	30	1,6
31		9,8			31	0,0			31	0,0	31	0,6			31	0,0			31	2,4
TOTAL		39,8		66,6		8,6		89,6		149,0		237,7		264,9		187,8		53,9		82,3
Daily average		1,28		2,22		0,27		2,98		4,80		7,66		9,46		6,05		1,79		2,65

Source: CECOI/ SEMASA

9.6 APPENDIX F - APPENDIX OF TARGET MOSQUITO COMPLAINTS CORRESPONDING TO THE PROPOSED RADIUS FOR THE MEDICAL SCHOOL (VALPARAÍSO) AND BOM PASTOR RESERVOIRS

Chart 12 - Record of complaints from residents of Santo André regarding the common mosquito via the PMSA's Service Information System within a radius of up to 2 km from the RM-09 Reservoirs at the Faculty of Medicine and the Bom Pastor Retention Reservoir between August 2014 and May 2015.

Tp. Doc	Doc. no.	Date	Official Neighbourhood	Prague	Results	Distance from the swimming pool
Tlm	2788220140	09/09/2014	Vila Valparaiso	Mosquitoes	PERNILONGO	1.3 kilometres
Tlm	2877820143	17/09/2014	Jardim Bela Vista	Mosquitoes	PERNILONGO	1.54 kilometres
Tlm	3309520139	08/08/2014	Vila Valparaiso	Mosquitoes	PERNILONGO	670 m
Tlm	3147520147	15/10/2014	Vila Valparaiso	Stilt	Stilt	620 m
Tlm	3251920148	27/10/2014	Vila Valparaiso	Aedes	Stilt	1 kilometre
Tlm	3219720147	23/10/2014	Vila Valparaiso	Stilt	Stilt	630 m
Tlm	3324720143	31/10/2014	Prince of Wales Village	Aedes	Stilt	770 m
Tlm	3346820140	03/11/2014	Vila Palmares	Stilt	Stilt	2 kilometres
Tlm	336742	05/11/20	Jardim Bom	Stilt	Stilt	11 m

	0141	14	Pastor			
Tlm	3337320149	03/11/2014	Vila Valparaiso	Stilt	Stilt	580 m
Tlm	3322420142	30/10/2014	Vila Floresta	Stilt	Stilt	400 m
Tlm	3378420148	06/11/2014	Jardim Bom Pastor	Stilt	Stilt	10 m
Tlm	3429020141	11/11/2014	Vila Floresta	Stilt	Stilt	10 m
Tlm	3456620146	13/11/2014	Jardim Bom Pastor	Stilt	Stilt	10 m
Tlm	3688920149	05/12/2014	Vila Valparaiso	Stilt	Stilt	310 m
Tlm	3834120148	22/12/2014	Jardim Bela Vista	Stilt	Stilt	1.93 kilometres
Tlm	3820220145	18/12/2014	Jardim Bom Pastor	Stilt	Stilt	920 m
Tlm	6920152	05/01/2015	Vila Guiomar	Stilt	Stilt	1.3 kilometres
e-mail	e-mail	09/01/2015	Vila Sacadura Cabral	Stilt	Stilt	1.1 kilometres
Tlm	127320156	15/01/2015	Vila Scarpelli	Stilt	Stilt	980 m
Tlm	191320153	20/01/2015	Vila Valparaiso	Stilt	Stilt	166 m
e-mail	e-mail	22/01/2015	Vila Sacadura Cabral	Stilt	Stilt	1.1 kilometres
Tlm	232620156	21/01/2015	Vila Gilda	Stilt	Stilt	1.19 kilometres
Tlm	335820158	28/01/2015	Vila Valparaiso	Stilt	Stilt	916 m
Tlm	328420158	28/01/2015	Vila Scarpelli	Stilt	Stilt	925 m
Tlm	359620159	29/01/2015	Vila Guiomar	Stilt	Stilt	1.5 kilometres
Tlm	416720153	30/01/2015	Jardim Bom Pastor	Stilt	Stilt	1 kilometre
Tlm	416120126	30/01/2015	Vila Gilda	Stilt	Stilt	1 kilometre
Tlm	395520150	30/01/2015	Jardim Bom Pastor	Stilt	Stilt	1 kilometre
Tlm	578720159	19/02/2015	Vila Guiomar	Bugs	Bugs	2 kilometres
Tlm	588120154	20/02/2015	Vila Palmares	Bugs	Bugs	2 kilometre

						s
Tlm	623820153	24/02/2015	Prince of Wales Village	Bugs	Bugs	830 m
Office	128/015	12/02/2015	Vila Guiomar	Bugs	Bugs	1.41 kilometres
Tlm	722220154	27/02/2015	Paradise	Bugs	Bugs	2.0 kilometres
e-mail	e-mail	01/03/2015	Vila Guiomar	Bugs	Bugs	1.65 kilometres
Tlm	769420150	04/03/2015	Jardim Bela Vista	Bugs	Bugs	1.5 kilometres
Tlm	871220156	12/03/2015	Vila Bastos	Mosquitoes	Bugs	1.2 kilometres
Tlm	928720152	18/03/2015	Jardim Bela Vista	Bugs	Bugs	1.42 kilometres
Report	report	17/03/2015	Vila Scarpelli	Bugs	Bugs	1.41 kilometres
Tlm	1121620156	31/03/2015	Vila Palmares	Bugs	Bugs	1, 120 km
Tlm	1174320157	06/04/2015	Prince of Wales Village	Bugs	Bugs	300 m

Source: ***ECRVAS/GCZ/DVS/SMS/PMSA***
Bom Pastor swimming pool coverage area
Valparaíso swimming pool coverage area

The following information has been omitted Service Order (SO) number; Requestor's name; Requestor's or Respondent's address; Service response.

9.7 APPENDIX G - PRODUCT

The products generated by this research are summarised below and can be incorporated into corporate management models. Suggestions for improvements arising from critical analysis and user experience are always welcome.

SUGGESTED APPLICATION IN DECISION-MAKING PROCESSES:

The distribution of the indices in the proposed matrices establishes a diagnosis of the municipality being assessed, but does not necessarily help in the process of defining/deciding on forms of intervention or when and where to proceed with intervention. It is necessary to establish the reference values for each index in order to enable a more appropriate assessment of each municipality, establishing its position in relation to its region, state or country.

One way of establishing reference values for the indices is, for example, to determine the median of the indicator values for a group of municipalities, areas or establishments, in this case two swimming pools. The median is the value that establishes, in a distribution, that half the values in the sample are above and half below this value.

Another essential step that must be taken and improved in practice is to analyse the historical series of infestations in each locality or municipality.

This way of establishing reference values helps in the decision-making process by allowing comparisons and, consequently, the selection of investment sites. It also has the advantage of working with the idea of making incremental qualitative leaps possible, even if the sustainability bias is not actually considered at first.

The median indices for the Santo André pools from August 2014 to May 2015 are shown in **Table 13** below:

Table 13 - Median values of infestation rates and complaints corresponding to the RM-9 Faculdade de Medicina (Valparaíso) and Bom Pastor reservoirs from August 2014 to May 20015.

	Valparaíso Swimming Pool	Bom Pastor Swimming Pool	General
DIL	0	0,1	0,15
IMAM	0,72	1,98	1,43
IMACR	0,2	0,1	0,45
IRC	1,5	0	1,5

By evaluating each municipality in relation to these figures, it is possible to establish a ranking of investment priorities.

This ranking or gradient is established as follows:

Light green colour - Situation under control, of little concern.

Yellow colour - Warning situation.

Orange colour - Critical situation.

Red colour - Alarming situation.

MANAGEMENT MEASURES ACCORDING TO THE CLASSIFIED SITUATION:

Situation under control - Educational measures and/or guidance are sufficient.

Cautionary situation - Mechanical or physical control measures are sufficient.

Critical situation - Chemical or biological control measures. Requires intervention, but is routine.

Alarming situation - Requires deployment of a team of operational controllers specifically for the problem, as well as mobilisation of the manager to involve other bodies or companies.

Therefore, the formulae proposed for calculating the indices for controlling Culex quinquefasciatus are as follows:

DIL (Density of Immatures per Litre):

$$DIL = \frac{\text{number of pupae plus number of larvae at stages L3 and L4}}{\text{1000 millilitres (1 litre)}}$$

Table 14 - Gradient values to be considered in relation to the Immature per Litre - DIL.

Gradient	DIL

Under Control	Up to 3.0
Attention	4,0 a 50,0
Critical	51,0 a 499,0
Alarming	> 500,0

IMAM (Index of Mosquitoes Aspirated per Minute):

$$IMAM = \frac{\text{number of } \textit{Culex} \text{ mosquitoes collected and identified during N}}{N}$$

Where N is the duration of minutes chosen for aspiration using the Nasci Aspirator. In the case of this study, N = 5 (minutes).

Table 15 - Gradient values to be considered in relation to the Mosquitoes Aspirated per Minute - IMAM.

Gradient	IMAM
Under Control	Up to 1.4
Attention	1,41 a 4,0
Critical	4,01 a 149,0
Alarming	> 150,0

IMACR (Index of Mosquitoes Aspirated from Restrooms):

$$IMACR = \frac{\text{number of } \textit{Culex} \text{ mosquitoes collected and identified}}{\text{Rest box}}$$

Mosquitoes will be collected using the Marucci Christmas Vacuum Cleaner until there are none left inside the Resting Box, giving the value in the numerator. The denominator should always be 1 (one).

In this work we used resting boxes measuring 36 centimetres high by 40 centimetres long and 28.5 centimetres wide, made of white styrofoam, painted internally with matt black latex paint, with a circular opening in the lid with a diameter of 15 centimetres, and the lid closed with adhesive tape using damp cotton cloth as an attractant. The box is exposed to the environment for approximately 18 hours, from the afternoon of the previous day until the morning of collection, when the opening is closed with sulphite paper and adhesive tape and the reading and identification is carried out after aspiration in the laboratory.

Table 16 - Gradient values to be considered in relation to the Mosquitoes Aspirated from Restrooms Index - IMACR.

Gradient	IMACR
Under Control	Up to 0.4
Attention	0,41 a 10,0
Critical	10,01 a 29,9
Alarming	> 30,0

IRC (Corresponding Complaint Index):

$$IRC = \frac{\text{number of complaints within a two-kilometre radius of the chosen breeding site}}{\text{One month period}}$$

The accounting of complaints can be checked to verify the target pest in question.

Table 17 - Gradient values to be considered in relation to the Corresponding Claims - IRC.

Gradient	IRC
Under Control	Up to 1.0
Attention	1,1 a 2,0
Critical	2,01 a 49,9
Alarming	> 50,0

We recommend setting up data recording spreadsheets (Excel programme) to monitor data for all the indices proposed here.

It is recommended that a historical series of Culex quinquefasciatus infestations be established in order to construct more reliable data for the location and time when applying these indices.

ANNEXES

ANNEX A - ANNEX OF RELATIVE HUMIDITY MEASURED IN THE REGION RELATING TO THE MEDICAL SCHOOL (VALPARAÍSO) AND BOM PASTOR RESERVOIRS

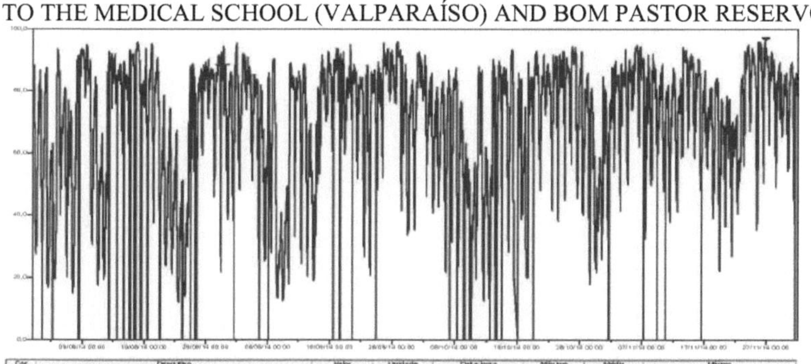

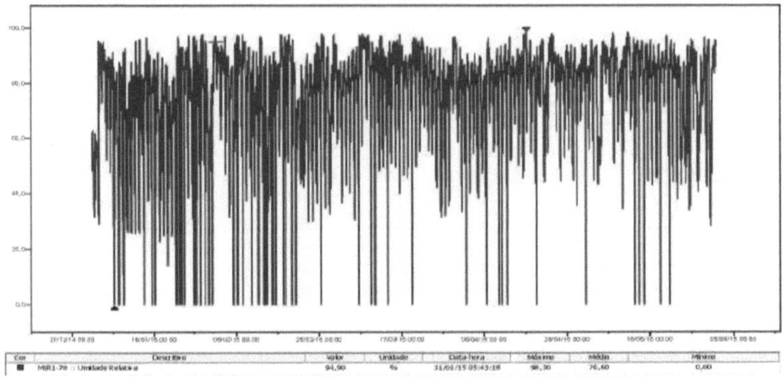

Source: CECOI/SEMASA/PMSA

ANNEX B - ANNEX OF AVERAGE DAILY TEMPERATURE MEASURED IN THE REGION FOR THE MEDICAL SCHOOL (VALPARAÍSO) AND BOM PASTOR RESERVOIRS

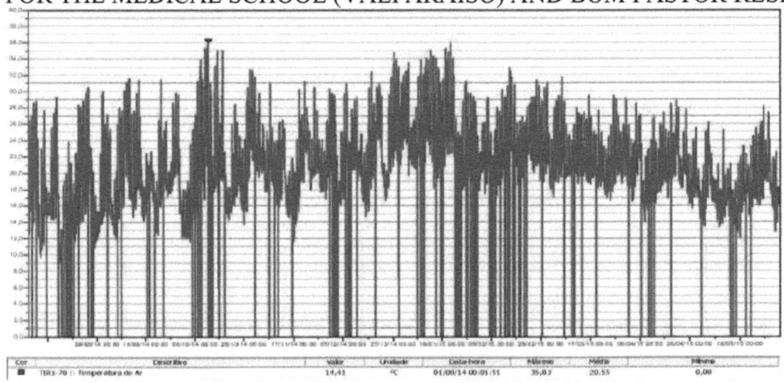

Source: CECOI/SEMASA/PMSA

MIX
Papier aus verantwortungsvollen Quellen
Paper from responsible sources
FSC® C105338

Printed by Books on Demand GmbH, Norderstedt / Germany